No. 2764
$15.95

FARMSTEAD MAGAZINE'S
GUIDE TO
ANIMAL
HUSBANDRY

EDITORS OF FARMSTEAD MAGAZINE

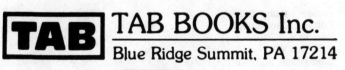

TAB BOOKS Inc.
Blue Ridge Summit, PA 17214

FIRST EDITION

FIRST PRINTING

Copyright © 1987 by TAB BOOKS Inc.

Printed in the United States of America

Reproduction or publication of the content in any manner, without express permission of the publisher, is prohibited. No liability is assumed with respect to the use of the information herein.

Library of Congress Cataloging in Publication Data

Farmstead magazine's guide to animal husbandry.

    Includes index.
    1. Livestock. 2. Bee culture. 3. Rabbits.
I. Farmstead magazine.
SF65.2.F38   1987   636   87-5045
ISBN 0-8306-9464-1
ISBN 0-8306-2764-2 (pbk.)

Questions regarding the content of this book
should be addressed to:

    Reader Inquiry Branch
    Editorial Department
    TAB BOOKS Inc.
    P.O. Box 40
    Blue Ridge Summit, PA 17214

Contents

Introduction

Homesteading is not only living off the land for self-sufficiency; it is incorporating new and old proven methods of raising livestock at the most economically devised means. Within these pages is a multitude of possibilities, from beekeeping or raising your own chickens to working the soil using oxen or horses. There are articles on how to buy a pig, cow, or horse, and what to look for when examining the animal. There are some homespun reminiscences, too.

This book is a sampling of some of the alternatives that are being used by others who want to rely more on themselves and the land to provide their food. All material contained in this work was made possible through the consent and cooperation of the editors of *Farmstead Magazine*.

Chapter 1

Beekeeping

For the beginner or those newly initiated to the hobby of beekeeping to be successful, you must do a fair amount of reading and pay particular attention to details. Beekeeping bulletins from the various states are available and provide specific details for your area.

A BEGINNER'S GUIDE TO RAISING BEES

Some people are allergic to bee stings and it is my opinion that anyone attempting to venture into the hobby should have a sensitivity test. This is a simple test done by your family doctor or local health clinic. Perhaps most of us have been stung by bees or wasps and realize the resulting effects. However, if you do not know, it is a good precaution to have this test prior to keeping bees, as violent reactions, even death, may result from bee stings.

Visiting a beekeeper for one day or talking with someone of notable experience, or even taking a beginner's course in beekeeping, is an excellent idea. However, everyone cannot benefit from such contacts, and, it is hoped that this article provides the guidance and information necessary.

Races or Varieties

In the United States we would not know commercial beekeeping as it now exists without the Italian race of honeybees. The

1

honeybee, *Apis mellifera*, has a number of races or varieties throughout the world; however, only four have been used in the United States with various degrees of success. When Europeans began establishing the 13 colonies in this country, they also brought the honeybee with them. The honeybee is often referred to in American Indian history as "the white man's fly."

At this time, the term *race* should be defined. It is used here with reference to any population of honeybees that has remained isolated long enough in a certain geological region to develop characteristics and qualities different from those of other populations in other geographical regions. For such characteristics to develop, it took millions of years and we have had bees in Maine for less than 400 years.

The three races commonly used in Maine are the Italian, its origin being from Italy; the Carniolan, from Central Europe and sometimes referred to as Carnica; and the Caucasian, from the Central Caucasus Mountains of Russia. These three races are available to most beekeepers.

It's advisable for the beginner to start with the race most commonly available, which will in all probability be the Italian. After experience, some extensive reading, and/or courses, you may wish to try other races. Also, two colonies is usually recommended for beginners, so that comparisons, combinations, or other manipulations might be made with some degree of success.

Siting of the Apiary

Proper siting of the apiary is crucial as it's one of the most important steps for proper overwintering. It is necessary that the apiary (and bees) receive the early morning rays of the sun (when it shines), so they get the maximum daily radiation, so it is essential that hives face the Southeast, with proper protection from the wind. Windbreaks such as hedgerows, pine trees, or stonewalls often serve as good protection from the North and West. The windbreak helps reduce the chill factor and allows the colony to maintain a uniform hive temperature. Hives in the North should never be shaded, as there are very few days of 90 °F. or greater that might affect the colony. Shade is necessary in places where summer temperatures commonly are above 90 °F. It's the winter months that are the greatest problem to beginners in the North. Therefore, a slight rise or slope of the land is helpful for good air drainage. This will aid in the removal of moisture during the winter months.

You do not have to plant a flower garden or forage for your bees unless you live in an area where flowering plants are few. Bees will range up to two miles, according to the literature, if foraging demands such flights. However, they normally fly within one mile or less of their hive. Bees like clover the best, or I should say you will get better yields of honey where clover blooms are common. Some parts of the U.S. provide various plant blooms and therefore it becomes a little bit dangerous to generalize. One thing is for sure, it won't pay you to buy equipment and plow land to raise forage crops for bees unless you're doing other associated farming to justify the expense. If that's the case, then back to the clovers,— alsike, white dutch, white sweet, and red top clovers have been used in various states with success. I would recommend you check with your county Extension agent if you do plant any specific bee flowers for foraging.

Biology of Bees

The behavior and relationships of bees have been studied for years and since the honeybee is such a beneficial insect, studies will undoubtedly continue. The honeybee is a social insect and lives as a group called a colony. They are unable to survive for long singularly and must group or cluster together into the colony.

Individuals of the Colony

There are three castes or individual bees that make up the normal colony or hive. The first and foremost is the *queen*. This is a fertilized female who is often considered the mother of the colony. She is the "egg factory" and her function is to lay 1500 to 2000 eggs per day, during the summer months. She is capable of laying fertilized and unfertilized eggs. She is not necessarily a slave to the workers, nor is she the ruler. However, her activities are governed by the workers. Reports in the literature have stated that a queen may live as long as seven years; however, under normal beekeeping management they are rarely kept more than three years. Some beekeepers purposely re-queen their colonies on a yearly basis to insure productive hives. Also, there is normally one queen per colony, but in some it is possible to find a virgin queen on the combs. When two queens come in contact they do mortal combat and the survivor will continue to be the egg producer for the colony.

The queen is the largest member of the hive and is easily distinguished from the workers and drones by her long body form. She has a short tongue and is fed by the workers who groom and care for her at all times. She has a smooth stinger with a large venom gland and apparently does not use her stinger unless on a rival queen.

The second type of bee in the colony is the *worker* and the bulk of the population is made of these sexually undeveloped females. They are able to lay eggs but the eggs develop into drones. Workers are smaller than the drones and queen; however, they perform all the working tasks of the colony. During the summer months in the North the worker population of a strong and healthy hive should be between 45,000-60,000.

Workers have longer tongues and pollen baskets on their hind legs for gathering pollen. They produce wax for comb-building, and they gather all the nectar which is converted into honey. It is little wonder, after performing so many tasks that this caste lives but three to six weeks during the summer months. The workers that hatch in early fall do survive through the winter months and are replaced in the spring when brood production commences.

It is necessary to have a high worker population during the nectar flows so that a crop of honey is made.

The worker is also the guardian of the hive. She is the individual with the stinger which does smart when inserted into any enemy's skin. The stinger is barbed and when it is inserted into its victim, the worker struggles to remove it; in doing so, she tears the venom sac and stinger from her abdomen and dies shortly after. Workers do not sting unless they are injured, disturbed, or in the process of defending the colony.

The third and last individual that is significant to the colony is the *drone*. This caste is the largest and stoutest member of the colony. The drone has two large compound eyes which also distinguish it from the others. These individuals live but a few days, and several hundred are tolerated during the colony reproduction and growing season. Drones are the fertilized males, and their apparent function is not to perform any other work except fertilizing virgin queens during the mating flights. Also, only one drone mates with the queen in these mating flights and the drone who is successful during the flight dies shortly after, since the mating organs are torn from the abdomen.

The drone has a short tongue and cannot feed himself so he gets his meals from foraging workers coming from the field. When the first cold frosty nights come in the fall workers carry drones from the hive, as they are no longer of any use to the colony's survival during the winter.

The three individual castes live and protect each other as a colony or unit. Individually they cannot survive, but as a colony they are "all for one and one for all." The queen produces the eggs, the workers perform the hive duties, and the drones fertilize new queens. Since this biological fact is so important, we will refer to the hive or colony from now on as the unit of production.

Development of Honeybees

Honeybees undergo a complete metamorphosis, which means four stages of development: egg, larva, pupa, and adult. The time required for each caste varies and in Table 1-1, this difference is shown.

Table 1-1. Number of Days Required for Caste Development In Honeybees.

Caste	Egg	Larva	Pupa	Adult
Queen	3	5 1/2	7 1/2	16
Worker	3	6	12	21
Drone	3	6 1/2	14 1/2	24

There is no difference in the time of development between fertilized and unfertilized eggs. However, there is a difference, obviously, in all subsequent stages. Studies have shown that queen determination is governed by moisture content of the larval food during the first day. Worker larvae receive more diluted food with higher moisture content than the queen during the first days of larval life. On the other hand, drones are determined by the size of the cell in which the queen lays the egg. There are five worker cells per square inch of comb but only four drone cells per square inch.

Queen cells are different from worker or drone cells, and are located at the bottoms of the frames or in the middle of the brood frames. These cells are pendent and long, and described as having the looks of a peanut shell.

Activities in the Hive

The division of labor in the honeybee colony is one area of interest that has produced a number of studies. A German scientist,

Rosch (1925), was the first to publish observations on the activities of bees within the hive. The workers go from cleaning brood cells, to keeping brood warm, feeding larvae and queen, producing wax for comb building, serving as guards, to foraging in the field for nectar, pollen, and water. This sequence is subject to modification any time an event occurs which interrupts their normal colony life.

When you look into a colony of bees for the first time, you may think there is complete chaos. Bees are running, crawling, and flying with what seems a lack of organization; however, there is actually a definite sequence of events based on the age of the bees. Such activities of the worker bees is divided into three categories: Nursing of the brood (eggs and larvae); work duties inside the hive; and foraging and field activities.

Getting Started

This is usually a problem for beginners because you will probably want a gentle bee, one that makes a lot of honey, does not require much handling, is disease-free, and doesn't swarm. Well, all those characteristics in one bee are not possible. However, there are four ways for beginners to obtain populations of bees:

● Order a three-pound package with a queen from a Southern bee breeder.

● Order a three or four-frame nuc from a Southern dealer or a local beekeeper who sells nucs, with a queen.

● Purchase a complete hive with bees and queen from a commercial beekeeper, or someone who is discontinuing beekeeping.

● Obtain a swarm or large cluster sometime in May that could be hived and manipulated.

The best time to start with bees is early in the spring and hopefully you will have disease-free bees in good healthy condition. Items being shipped to Northern States will be inspected for disease but bees sold in state should have the approval of the State Bee Inspector. To find out who this person is, call your State Department of Agriculture, Division of Plant Industry of Apiculture.

For the beginner, I recommend the purchase of three-pound packages with queen. This allows the newly initiated to see the marked queen-egg-laying pattern develop and observe hive growth.

The beginner is faced with about a $200.00 expense to start, with one hive, bees and related equipment. This does not include an extractor, capping knife, jars, and so forth. There is no annual maintenance cost unless you wish to expand your colony numbers. The $200.00 figure may be cut in half if you are an innovative do-it-yourselfer, or a carpenter. Then your cost might be in the order of $100.00. One thing I recommend beginners to avoid is used equipment. You might pick up some bee diseases in old equipment that has been around for years. Therefore, have your state Bee Inspector advise you prior to your purchase. There are many dealers and distributors of bee supplies in the U.S. and there is nothing like new equipment that fits together and works well. Some of the larger dealers whom you may wish to write to for free catalogs are listed below:

Dadant & Sons, Inc.
Hamilton, Illinois, 62341

A.I. Root Co.
Medina, Ohio, 44256

The Walter T. Kelley Co.
Clarkson, Kentucky, 42726

Hubbard Apiaries, Inc.
Onsted, Michigan, 49265

These companies sometimes have dealers in your state through which you may purchase supplies or you may deal directly with the company.

Langstroth Hive

Thanks to L.L. Langstroth, who developed the movable frame hive and utilized an earlier discovery called bee space, we now have a convenient and manageable home for honeybees. His 10-frame beehive, sometimes called the Langstroth-style hive, has become the standard hive in the United States. In the North, most hobby beekeepers utilize the two hive body system with supers added above the queen excluder.

In discussing the various parts, we will work from the bottom board to the top cover. The bottom board is where the first hive body, or brood chamber, sits. It is where the brood rearing takes place. This provides room for winter stores and adequate space for brood rearing and colony expansion. There are some beekeepers

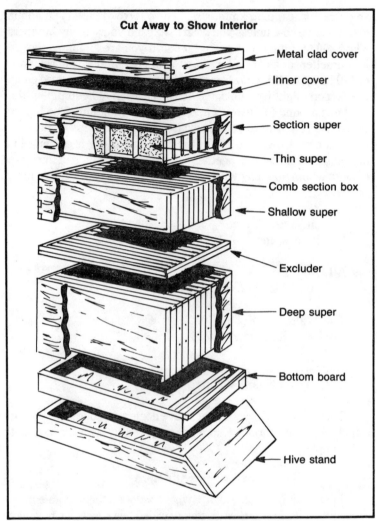

Cut Away to Show Interior

Metal clad cover

Inner cover

Section super

Thin super

Comb section box

Shallow super

Excluder

Deep super

Bottom board

Hive stand

Fig. 1-1. An exploded view of a modern bee hive.

who use all hive bodies, even for honey production.

This practice is not recommended for beginners or the elderly. The weight of one full hive body or deep super is well over 50 pounds. This is a significant disadvantage to most beekeepers when manipulation of the deep supers is required. The next item is the queen excluder, which keeps the queen in the brood chambers or hive body during her egg-laying activities. Some beekeepers, myself in particular, feel that queen excluders slow down the movement of workers since they have to pass through the wire screen while working between the hive body and supers. However, if you are going to have comb honey, then a queen excluder is a must. The supers, shallow type, are next, and when these are full they weigh about 30 pounds. Many beekeepers utilize nine frames or less in the shallow supers. This allows for good comb production with well-sealed caps that are easily removed during extraction.

Above the supers is the inner cover with a central opening used for a bee escape or feeding purposes. Last is the outer cover, usually made of a metal coating over wood.

Each of the hive bodies contains 10 frames with wax foundation or drawn comb for brood production. Each super should have nine frames or less as explained above. The design of this equipment has changed very little over the past 50 years; however, materials such as plastics are being used for frames, foundation and hive bodies. All of these materials may be obtained from a beekeeping supply catalog.

Installing Package Bees

The best time to have package bees arrive in Northern states is the last of April, or the first week of May. This, of course, varies with each season; however feeding of package bees is recommended so that if inclement weather arrives the bees will have food available.

After your bees arrive, and you have installed them into their new home, you should leave them alone for about one week. This allows them time to become organized, for the queen to start laying, and for them to become acclimated to their surroundings. One thing I should point out is a safe method for releasing your new queen, which will be in a small cage. There will be a plug with candy (white sugar) in one end; push a small nail hole in the candy after removing the cork plug and put the cage in the hive between two frames. Within one week the queen will be released and hopefully you will be in business. Honey flows or more properly called, nectar flows, vary in each state depending on the flowers in bloom; but in most northern latitudes the first "honey flow" will last through June depending on weather conditions. Then a second flow occurs from August to October. You should tend to your bees at least once every two weeks to check on disease problems, progress of brood rearing and swarming conditions, adding honey supers when necessary, and all sorts of associated matters to the husbandry of beekeeping, if you are to have a crop.

Protective items are necessary when it comes to handling bees, and you should wear gloves, a veil to protect the face, rubber bands around your trousers to protect your legs, and a smoker to quiet down the bees during handling. A good hive tool plus a bee brush are also necessary in the manipulation of hive frames. The hive tool is used as a pry to separate the frames and hive bodies; the bee brush is to move clusters of bees by gentle brush motions which will not excite the hive. Some beekeepers prefer a full bee suit with a zipper veil. Clothing such as black colors, wool, or fuzzy materials are irritants to bees and they often attack such clothing. White is the preferred color to wear while working with bees, followed by tan.

The smoker is used to quiet the bees prior to opening of the hive. Use very little, otherwise you will only anger them and prevent proper handling. Never stand in front of the hive entrance because bees will be hitting you as they approach their hive. Blow a little smoke in the entrance and then a little under the inner cover. Wait a few seconds and the bees should be quiet enough so that you can open the inner cover and handle them. The theory behind using smoke is varied, but it causes the bees to go down between the frames, and fill up with honey. Some beekeepers feel that the bees fear that the hive is on fire. Avoid too much smoke which might make the bees roar, and avoid any rough treatment that might anger them. Move slowly, avoid killing them, and you will be surprised as to what you can do as time goes by.

Smoker fuel varies with beekeepers; staghorn sumac, burlap, corn cobs, and bailing twine have been used with success. Never use soiled or oily rags or cotton clothing with dyes. The type of smoke is a cool white nontoxic odor which will cause the bees to remain in a "good mood."

Harvesting Honey

When harvest time comes, this means taking the crop or surplus of honey, while still leaving enough for the bees to have during the winter. Most beekeepers winter their bees in two brood chambers, or sometimes it's called two hive bodies, and the extra honey is in the supers. Two hive bodies with bees and honey that weigh over 100 lbs. will provide adequate winter stores. The next thing is, how do we get the honey out of the frames in the supers? This is easier said than done. What you need are some jars, an uncapping knife, an extractor or centrifuge (homemade types can be fabricated), and lots of time and patience. The honey can be heated and filtered; however, my 16 years of beekeeping has told me not to filter or heat unless it's for the roadside stand. Most people like to see honey very clear. To do so means heating, filtering, and losing some of the valuable enzymes, flavor and food value. Therefore, if you like "raw honey," you are in for a treat. The pieces of wax, propolis, and even dead bees are easily picked out when you fill your jars. Remember this: there is no known active bacteria harmful to man which can live in a low moisture environment that pure honey provides.

Extracting honey is a task too complicated to cover thoroughly in one article, so again, I want to emphasize how important it is for the beginner to read and spend time at the hobby. I remember when I started, I spent nearly one year reviewing literature prior to jumping into the hobby. In fact, it's good to start during the fall and winter months, in my opinion, because a course on beekeeping might be available to you in your state before you start beekeeping next spring.

GETTING A LINE ON BEES

Sitting in a field of goldenrod, early, on a hot, dry August day, you gaze at the sky. You watch intently as a minute insect circles a box in the field several times and flies straight into the woods. You are beginning the time-consuming, frustrating, but fascinating bee hunt.

The bee hunt and taking up of swarms of thousands of wild honey bees is an activity known to few newcomers to country living. A bee hunt may take only a day or may last a week or longer. You may be rewarded with a hundred pounds of honey, or you may get only the amount you fed the bees you were following . Whether you want the nutritious, golden honey or the exercise and adventure, a bee hunt will cost very little but your time and energy.

A local farmer, asked how to hunt bees, replied, "Why, just put chalk on them and track them through the woods!" Pulling your leg? Not really. There are many ways to hunt bees. I will describe a method which has been successful for me.

First find out who keeps bees in the area you want to hunt. Hunt at least two miles from any domestic hive, since bees generally feed within a mile of their hive. Start in a large open field where bees are gathering nectar from goldenrod or other fall flowers. Bee hunting is generally easier in the late summer and fall than at other seasons because bees are working very hard to add to their stores.

Basic equipment for the hunt is a compass, watch, powdered colored chalk, two four-inch square feeding boxes on three feet high poles sharpened on one end and a four-inch square catching box. The catching box has a glass top, an open bottom and a sliding panel in the middle; or you can just use a tumbler and piece of cardboard. If you are allergic to bee stings, wear protective clothing.

Inside the feeding boxes place a piece of honey comb and partially fill this with a sticky, runny mixture of sugar and water. A few drops of anisette can be added to create a stronger aroma. Honey may be used but must be diluted with hot water.

The nectar bees collect from flowers is very thin and they can carry about one-fourth of their own weight. If you use too thick and heavy a mixture, the bees will pick up too much weight and wobble in flight preventing you from getting a true line on them. Or they may need to stop for a rest on the way to their hive, and you will be unable to figure how far away the hive is.

Use the catching box to move bees off the flowers and into the feeding box to start the hunt. Capture a bee while it is collecting nectar from the goldenrod. When it flies to the glass top, close the sliding panel. After getting eight to ten bees, release them into the feeding box. Use this same procedure to move the bees each time you move the feeding box.

When a bee has collected enough sugar water, it will spiral upward marking the location of the box and fly straight to the hive. After a number of bees have started working, mark three or four with different colors of powdered chalk. Simply dip a twig or piece of grass into the wet, pasty chalk and touch it to the rear of the bee. Don't worry about being stung as bees generally won't sting while they are working, only when something is taken from them.

Watch the bees as they spiral above the box; when they start a straight flight, take a line of them with your compass. If you are lucky during this hunt, the bees you have marked will not be eaten in flight by birds. If bees are going off in two different directions from your feeding box, is usually means they are from two different hives. You must select one of these directions to continue your bee hunt. You can try to find the other hive later.

The distance to the hive can be estimated using a formula given in the ABC-XYZ's of Beekeeping. A bee will average a mile in five minutes and spend about two minutes in the hive. Measure the time that some of the colored bees are gone from the feeding box in minutes. Subtract two minutes from this time and divide the remainder by ten. This will give you the approximate distance to the hive in miles. Remember this is only an approximation because bees might travel faster than this in clearings or slower in thick woods. Wind speed also affects a bee's flight speed.

After you have determined the direction of the bees' flight, set the second feeding box on the line of flight in as large a clearing as possible. If this location is in the woods, move the box only a short distance, no more than 200 feet, in order to get a good line on the bees.

Hazards such as thick woods, swamps and fields of cows make this part of the hunt very time-consuming. It has taken me as long as a week to locate a hive of bees only twenty minutes away from the field of goldenrod. If you are caught in a thunderstorm, you will have to postpone the hunt until it clears. Continue jumping the bees this way until they are only gone three minutes. By then you'll know the hive is within a couple of hundred yards.

In northern New England the hive is often located in a hardwood tree with the entrance on the southern side. The entrances range from large hollows to small knotholes. If you cannot locate the hive when you know you are close, start a line from another direction. The hive should be at the junction of the two lines.

When you locate the hive, check to see if there are any initials on the tree. Someone may already have claimed the hive; if not, claim it by marking your initials there.

Now, if you are into bee hunting for more than just the sport, it is time to collect the bees and honey. If you are not experienced, I would suggest that you seek the help of an experienced beekeeper. If the tree is not valuable and the landowner gives permission, it is easier to collect the hive after cutting the tree.

With as many as 25,000 to 75,000 angry bees around, you must wear protective clothing—gloves, tied pant legs, tight sleeves and a hat with a veil. You will need a smoker, an ax, a saw, a pickup box and a bucket for the honey.

Cut the tree to fall into smaller brush to prevent it from smashing on the ground. You should smoke the entrance to the hive as soon as possible after the tree is felled in order to calm the bees. Now see how large the hollow is by tapping on the tree with an ax. Make saw cuts across the hollow for the entire length of it without cutting into the hollow if possible. Split out these sections with the ax and remove them. The remainder should resemble a canoe.

You will see that some of the layers of comb are very dark and some light. The darker the section, the older the comb. Remove the darker layers and any layers that contain brood and place them in the pickup box next to the tree. Take the remaining layers with you in the bucket.

Using the smoker you will be able to force the bees into a group which you can scoop up with your gloves and place into the box. It is essential to capture the queen bee which is larger and more brightly colored than the other bees. If you do not see the queen, which is the usual case, you can tell if she is in the box by the behavior of the bees. If they tend to stay in the box the queen is probably there. If they tend to dart out of the box and back into the hive, it means she is probably still in the hive. In that case, you must find her and get her in the box.

Bees are notorious robbers, and within a few days, the worker bees will transfer most of the honey from the hive to the box. They will also remove all of the dead bees from the box which were killed when you were working. After the bees have had time to transfer all the honey, you can return late in the evening when the bees are inactive, screen the opening and take the bees home.

You have now experienced a bee hunt, and if you were fortunate and persistent, you have been rewarded with an ample supply of honey and a source for years to come.

FIND A HONEYBEE SWARM

Most folks run indoors and call for help when a cloud of honeybees from a overcrowded hive swarms in on their property. This usually occurs in the Spring. They are a formidable sight, but when you get to know bees, you'll realize that at this time, they are the most gentle and easy to approach.

If you decide to do as I did, and send notices out to local papers, fire and police departments that you provide free and efficient removal of swarms, you need to know what to do when the call comes that a swarm is in the area. The first thing you do is get the address and phone number of the people who called for help. Call them and get the following information: Do you need a ladder to reach them? Are they hanging in a cluster, or flying around aimlessly, but within a definite objective area, such as a tree-limb or bush? Will you be allowed to cut a branch or two to get them? The only swarms I bother with are those that can be reached with a 20-foot extension ladder, or those lower to the ground, and where the branch they're on can be cut or is small enough in diameter to be "snapped" sharply in order to dislodge the bees clinging to it in a cluster.

Two or three frames from a hive with foundation in them (see drawing) are placed in the center of a deep super. This will attract the bees later, when they are shaken into the box that will be their new home. But first, nail or staple a bottom board onto the super, as this will be the bottom section of your new hive when you set it onto its stand. Fill the entrance with grass, packed in tightly so that no bees can get through the opening. (When the grass dries, the bees will remove it to provide their own exit.) Set the telescoping outer cover over the whole affair, pack it up, and get out to the site where your swarm of bees are making a nuisance of themselves. (Oh, and if you'd like to smear a little honey on the foundation before you leave, go ahead! If you have drawn-out comb, use it instead of new foundation.)

Double check that your bee-veil is tucked in all around, that your gloves cover the openings in your sleeves, and that your pantlegs are tucked into your stockings. Never mind what people will say, they'll think you're crazy for being near those bees in the first place. Take a deep breath, compose yourself, and one way or another, hold the box from 6-12 inches below the cluster. The objective is to get as many bees as possible into the box on the first try. The queen is in the center of that mass of bees, and if you get her in there, you'll know by the way the rest of them settle down into the box with her.

If you've decided to cut the branch, cut away the excess on the end of the branch furthest from the trunk, and when you make your second cut on the other side of the swarm to remove it, you won't have a large, off-balance limb to handle, but a short stubby piece of wood full of bees. Put the whole thing in the hive, branch and all, and close the cover. Since you've filled the entrance with grass, they won't be able to get out. Load them up, thank the folks, (most likely, they'll thank you!) and get on home.

Set your new hive up as close as possible to where you plan to keep it permanently, about 6-12 inches off the ground, perhaps on cement blocks, and tilted slightly forward to allow moisture to run out, and to avoid letting rain in. Put in the other eight frames the next day, when your bees have had time to get accustomed to their new home. Over the box that they are in, place an empty shallow super. Take an 18-ounce peanut butter jar and fill it with a mixture of sugar and water (50-50). Cover it with a lid punctured with small nailholes. Place the jar upside down over the frames, and let it drip down. This will keep the bees fed until they are able to locate a source of nectar. Remember. Your new bees must be checked for diseases periodically.

Voila! You are now a beekeeper, right? Not quite, You're well on your way, but there's a lot to beekeeping. Trial and error, much reading and learning, and talking to experienced beemen will help a lot. Just remember that you can keep your bees healthy and happy, and producing honey, by doing your job properly. They became your responsibility when you captured them, and deserve your attention and care.

WILD BEE HUNTER

Seventy years ago, as a school boy of fifteen, I was an ardent bee hunter. Along with two boon companions, we tramped the fields and pastures of Northwestern Connecticut, hoping to start a beeline to a honey tree.

Before our search began, we had to obtain a small box to convert to our special type of bee box. Not any old box would do. We knew just where to obtain one, so our plans started in the fall with the opening of school. At that time, the chalk the teachers used came in a small box about four inches by four inches by six inches. It was made of a light, odorless wood, was nicely dovetailed and had a sliding wood cover. It was that cover that made it so desirable.

We three were in different grades in school, so we had three chances to get one. By being the teacher's pet (bringing her red apples), and by staying after school to wash the blackboards and dust the erasers, we made sure that the chalk would be all used up by spring. In that way, we were almost sure of at least one box when we needed it.

Once in our hands, the transformation began. The sliding cover was replaced by a piece of window glass. My father was a carpenter and owned a glass cutter, so it was up to me to purloin a pane of glass from the barn or chicken coop and cut it down to size. Once that was done, a hole about the size of a quarter was cut in the end that wouldn't interfere with the glass slide. A trap door was made from the discarded wooden cover to fit over that hole. Next on the program was to persuade one of our mothers to buy a comb of honey from the grocery store and to let us have a good healthy hunk of it. With that in the box, and a ten cent bottle of oil of anise purchased from the village drug store with the pennies we had saved, we were ready for our trip to the fields.

The season started in spring when the first of the chestnut blossoms appeared (that was before the blight had wiped them out), and continued on until the first frost. The operation was simple.

Finding a bee working on a flower, we would open the trap door, and with the box in one hand and our cloth cap or a soft rag in the other, we would trap the bee by forcing it into the box, being careful not to injure the insect. The trap door was then closed and the cloth placed on top to darken the interior. After a couple of minutes, we would peek in to see if the bee was down working on the honey. If it was, the glass cover was removed, some of the oil of anise was dribbled around the edge of the box and it was put on top of a boulder or fence post where we could lie on the ground to watch which way the bee flew when it left. When that happened, the bee flew straight up in the air, circled about three times and then headed for home.

The theory of the oil of anise was that the odor helped the returning bee locate the source of the nectar. Whether it actually worked, we never knew, but it was part of the ritual. When the bee returned, it was often accompanied by another, so we got an idea of the direction they were traveling, and from the elapsed time, some idea of how far away their home was. After about their third or fourth return, and as soon as they were down on the honey, we would pick up the box and work our way in that direction. As soon as we saw that they were ready to take off, we would place the box on another high spot, and in that way, gradually approach the tree. Soon we would have a steady flow, and could pick up the box and walk without fear of losing the line. Once we found the tree, we would mark it. Then, shaking what bees we could out of the box, we replaced the glass cover, and with the trap door open to let the trapped bee escape, we left the area.

It wasn't always that easy. Sometimes the original bee never returned, or we would get tired of waiting and go somewhere else. A line could take you for a mile across country through brier patches and swamps, only to end up by someone's private bee hive. There weren't many farmers with hives, so we soon recognized when a line was headed their way. We hunted the edges of old burns and logged-off land since the old snags were good possibilities for a bee tree. Once a tree was found, it was marked and left until cold weather. With the help of older members of our families, it was cut. Sometimes it was worthwhile, and at others, a total loss. Once we worked all morning cutting down an old snag only to have it burst when it hit the ground, revealing it to be full of dying bees. The honey was black and foul-smelling, caused by foul brood disease.

Our most spectacular find came when we followed a line into

an old abandoned farm house. There was a story connected with the old place. Years before, a man had lived there alone and had built his house on the edge of a small, five acre swamp pond. He lived there several years and then mysteriously disappeared. A neighboring farmer chanced to stop by and found the door open and everything in order, but no trace of the owner. At that time, the swamp had been gradually taking over the pond and there were several floating islands throughout. It was thought that the man had ventured out on one of these and had broken through and drowned. No trace of him was ever found, and time and vandals wrecked the old house. White birches and sumac took over the fields. A few of the old apple trees remained, but outside of an occasional partridge hunter, few people ever went near the place.

When we arrived at the house, we could see that there was an old and large bee colony, judging from the large number of bees flying in and out of a knot hole in the old clapboards. We went home and told our folks. A couple of older brothers went back, looked it over, and decided to keep a close watch until cold weather. When the old plaster and lathing was torn out, the area between the studs was solid with comb. I remember helping to carry a wash boiler and several pails full of honey about half a mile to the nearest road because the old original road into the place had been abandoned years before. You can bet that everybody had honey on their pancakes that winter! Our mothers boiled the empty comb in water, skimming off the wax. Then they made little patties of it and covered them with cloth. These were used on the "smoothing irons" the women used to iron their shirtwaists and lace collars. I remember the cloth on my mother's was blue and white polka dot. Every farmer had a cake of wax to be used in pointing the thread used on his harness repairs.

One thing about bee hunting was that you learned your area like the palm of your hand. Tramping over the countryside was a school by itself. Your countryman's eye noticed where the best berry patches were, where the huckleberries grew, where the nut trees were for fall hunting. The baby-like footprints in the muddy bank of a brook told of raccoons, and the empty shells of the fresh water clams on a partially submerged rock or stump, spoke of muskrats nearby. The holes dug along an old line fence showed where skunks or an old dog fox had hunted for field mice, and the old grapevines and wild apple trees promised partridge and rabbit later on.

The season for nectar ended with the little blue flowers of the

Fig. 1-2. The parts of a Feldhake box: (a) observation windows; (b) sliding portion; (c) capturing room; (d) retaining room.

wild aster—those golden days of fall when the migrating bluebirds called their "cherry ripe" from the orchards, the robins spattered the stones of the hillside purple from their feast of ripe pokeberries, and nearly every mullen stalk had a wild canary prospecting for seeds. Those days are still fresh in my memory after all these years. Those were the happy days.

BUILD A BEE BOX

When sugar prices rose several years ago, natural food lovers and penny-wise housewives searched for an alternative sweetener. They found it in honey—the valuable product of a busy hive of honeybees. But, buying honey can become expensive, and the cost out-lay needed to buy new bee equipment and stock hives can be a souring experience.

Bees can be ordered from a supplier, but how much more fun and less expensive to go bee hunting and get bees for free! Well, almost. The only thing invested will be your time and the bait it takes to attract them.

Before going bee hunting, it's a good idea to know a little bit about these insects. They are gentle critters who go about their business and do not threaten to sting unless they feel threatened. A bee almost always stings out of self-defense. Fast, swatting motions can drive bees into attacking.

When they do get stung, most people incorrectly remove the bee stinger and heap misery on themselves in the process. Bee stingers hardly ever cause serious trouble if removed correctly.

The stinger is inserted into the flesh by the bee. Attached to the stinger is a sac, which upon close inspection appears whitish-beige in color. This sac contains the bee venom. By first scraping and lifting away the sac from the skin with the fingernail, the stinger may then be pulled out of the flesh and allowed to fall harmlessly to the ground.

This method of stinger removal does away with most of the pain of the stings and works to lessen fear while working around bees. However, when the stinger is squeezed, the poison sac ruptures and spills venom into the flesh, causing a painful wound.

Honeybees may be found quite readily in the woods, in town working around gardens, flowers and wild berry plants. Just any "bee" won't fill the bill. Only the honeybee is worth capturing. Trapped hornets, yellow jackets and bumble bees will only get you a hard time and possible stings.

Most wild flowers produce some kind of nectar that will attract honeybees. Because the flowers and weeds bloom at different times of the year, one may know almost exactly where to look for bees, by the season.

In the spring bees may be found by blooming fruit trees and houseyards and the woods. They may be located on dandelion and wild berry plants as well as on spring flowers. Summer shifts the bees to clover fields, patches of buckwheat, later blooming wild flowers and the garden variety flowers. In fall, the bees seek out goldenrod and very late blooming flowers and weeds for their nectar. By taking a moment to note what is in bloom a minimum of time will be spent locating honeybees industriously gathering nectar to take back to the hive.

Most methods of hunting any game require something with which to bag the quarry, and bee hunting is no exception. However, it costs a little to begin.

A bee box is necessary equipment and may be made quite easily even by an inexperienced woodcrafter. It need not be fancy nor even made out of high quality wood. The basic requirement is that it be approximately eight inches long and four inches wide. The bee box should have two windows at each end of the box, inlaid with clear glass, and a shutter over the window to let in light or block it out.

The box is divided in half with a partition. Approximately, the top two-thirds is stationary, and the bottom one-third a movable slide. The partition slide is the only way that bees are able to get in or out of the retaining compartment. The other half of the box has a lid mounted on hinges, with a shuttered window that may be lifted back.

Once the bee hunter has located an unwary honeybee that is busily gathering nectar, the box (with the lid open) is drawn quickly up underneath the plant where the bee is working. The bee should fall into the box, then the hunter should pull back the partition-slide and open the window shutter to the retaining compartment.

True to nature, the bee will fly under the partition, attracted to the light in the belief that it is an escape route. As soon as the bee can be viewed in the retaining compartment, shut the partition slide. Then, repeat the procedure with several subsequent bees. After about half a dozen bees have been captured, the bee hunter is ready to set about locating the bee tree.

Place the bee box on top of a fence post, boulder or other solid stationary object. Open the lidded compartment and place a piece of empty honeycomb (obtainable from a beekeeper or supply house) in the compartment. With a clean eyedropper, fill the comb with bait made from one part sugar to four parts water. If unable to locate and obtain empty honeycomb—a piece of clean sponge may be substituted—although it is not as desirable as real honeycomb.

After the comb is filled with the bait, the lid to the compartment should be closed and the top window shutter left open. Then open the partition slide so that the bees will leave the retaining room and go into the compartment with the sugar water. They should fly straight to it, and gorge themselves.

When the bees are filled, pull back the lid to free them, and step back. The bees will usually come and go to the bait box several times, hovering while they get their bearings, then fix the location of the box. Then they will head to their hive. When the news of easy syrup spreads throughout the bee colony, the number of hives working on the bait and forming a line will slowly increase.

After the original bees have left, transfer some bait into a container that closes easily, such as a cigar box. Trap several worker bees in it and move it gradually in the direction of the departure path of the bees. While a bee box with compartments is dandy to have if you bee hunt regularly, a cigar box will suffice for the initial trapping, but makes it more difficult to trap any number of bees without a few escaping.

The length of each move you make with the box is dictated by the terrain. In a field it can be quite far—perhaps as much as 200 yards. In the woods, however, a move of several yards may be the limit to prevent passing up the bee tree. Never pass up any tree without first checking it out.

During especially heavy nectar flows in some plants the bees will ignore the bait and not return once released. If this is the case, the hunter may as well consider the bee hunt officially over for about two weeks until the nectar flow has diminished and bees will again show interest in sugar water bait.

When trapping bees several times in the same general location, it is not unusual to capture bees working for two separate colonies. This is evident when bees gang up on each other, and the bees from the lesser represented colony are killed off by the bees in majority. Therefore, bee trees cannot be located simply by starting two groups of bees and hoping that the point of juncture in the line of bees will cross at the bee tree. This is not the case. They are often foraging for two separate colonies.

Most bees cease working in late afternoon and remain in the hive, so at that time, there are fewer bees flying around. If the bee hunt is to be continued the following day, the comb may be refilled and left in place with the cigar box being weighted with a rock or heavy object. This saves a lot of backtracking the next day.

When at long last the goal has been reached and the bee tree found, it is wise for the novice bee hunter to look up an experienced bee handler to help with the removal of the bees and honey. However, it is a good idea to look up the owner of the land and find out if he has any objection to the removal of the hollow bee tree. Cutting down a bee tree without permission can result in heavy fines. Usually, a share in the honey will alleviate any misgivings the landowner may have.

If you're interested in developing a thriving apiary, during swarming season in the later summer, in town or in the country, bees may find you without you hunting them down. Keep people alert to your willingness to remove bees, and they become a windfall to fill your hives.

Before setting out on a bee hunt, be advised, however, that the bee tree will probably not be found in a matter of minutes. It may take hours or even days, if the bee tree is far enough away, or craftily concealed. But any way that the bee hunter goes about it—for fun, profit, honey or for the bees, you've got to agree: bee hunting is one honey of an idea!

A BEE FOR THE NORTH

In numerous discussions I have had with Maine beekeepers, it is quite evident that there are two major problems with keeping bees in Maine. First and foremost, the problems of overwintering colonies, and second, the problem of low average yields, particularly in coastal areas. While little can be done to affect changes in the environment our bees are placed in, there are ways to increase both the average yield and reduce the number of colonies lost to the long severe Maine winter. In discussing these two problems, a general review of the average characteristics of the three races of bees of commercial importance is in order. Keep in mind that there are strains in each race that deviate from the average. Also, most of the stock available in the United States has been modified by cross-breeding, inbreeding, hybrid crosses, etc. The following is a listing of the three races as found in their native homeland and the advantages and disadvantages of each.

The Italian: Latin name *Apis Mellifera Ligustica*, original homeland—Italy. Italian bees are somewhat smaller than the German Bee (which are now extinct in the U.S.) with a slender abdomen and relatively long tongues. The color of the chitin (the outer covering of insects and crustacea) of the abdomen is brightened on the sterna as well as on the first two to four terga (yellow bands on their front parts). Also, the hairs of the Italians have yellowish color. This is particularly distinct on the drones.

Italians are usually calm and gentle, although this trait can be variable. The Italians have an extraordinarily strong disposition to breeding. The colonies start to breed early in spring and maintain a large brood area, regardless of honey flows, until late fall. This is called the "Mediterranean brood cycle," a breed characteristic which I consider detrimental here in Maine because of frequent failure of the honey flow. Exceptionally strong colonies occur with this brood cycle, and they show good performance, especially in midsummer. Unfortunately, here in Maine this coincides with the end of the summer flow. Despite the strong brood rearing trait, the

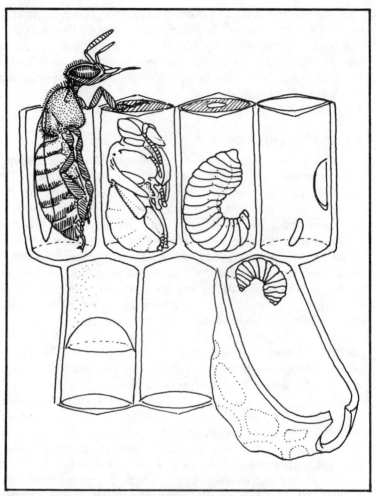

Fig. 1-3. Development of a honeybee.

Italians have little inclination to swarm. Thus, the overwintering of their strong colonies causes high consumption of food. In northern latitudes, this creates difficulties. There is a loss of worker bees due to early brood rearing (spring dwindling). Thus, colony development in spring will be slow and tardy, in case of a poor honey flow in summer, a shortage of food occurs because of the high food consumption.

The long tongue of the Italian makes the utilization of red clover feasible. The good building instinct of this race has been praised many times. Italians cap the honey with white cappings. However, the Italian is a bee of the Mediterranean climate, which is characterized by short, mild humid winters and dry summers with long honey flows, in contrast to Maine's short honey flows. The Italian bee has proved excellent in climates similar to that of the Mediterranean area, but for longer winters and tardy springs, with many relapses, it has no defense. It always has been difficult for southern races to settle down in the north. This can be readily seen in Europe. Many unsuccessful attempts were made with Italian bees north of the Alps. For more than 100 years they have been imported again and again, but they have never been able to establish themselves in good fashion.

Italian bees are also notorious for robbing other hives. If you open a new colony during a poor honey flow, the first yellow explorers will appear there quite soon. The sense of orientation of the Italian is weak. Often the bees fly to the wrong hive in a closely spaced apiary.

The Caucasian: Latin name *Apis mellifera caucasia*, original homeland—the high valleys of the central Caucasus. The shape and size of the body and hairs is very much like those of the carniolan (described later). The color of the chitin is dark, but frequently brown spots appear on the first bands of the abdomen. The hairs of the caucasian workers are lead gray. The hairs on the thorax of the drones are black. They have very long tongues, and no uniform color in their homeland, and the pure gray bee is more an ideal of breeders than a reality in nature.

Gentleness and calmness on the combs are the characteristics commonly emphasized for caucasians. They exhibit ardent brood production and raise strong colonies. However, they do not reach full strength before midsummer. The disposition to swarm is weak. They are very great users of propolis. In some strains this trait is so pronounced that they will seal their entrance with a curtain of propolis leaving only a small hole. This trait is very undesirable because it creates difficulty in manipulating hive equipment.

Overwintering in northern regions is not as good as would be expected because of their susceptibility to "Nosema," a disease of the adult bee. The performance of caucasians in red clover is not up to expectations based on the length of their tongues. The cappings of the honey are flat and dark, and the bees are inclined to drifting and robbing.

The Carniolans: Latin name *Apis mellifera carnica*, original homeland—Southern part of the Austrian Alps and the North Balkan (Yugoslavia), called Carnica in a broader sense, and Macedonia, including the entire Danube Valley (Hungary, Rumania and the Bulgaria).

Carniolans are generally quite similar to the Italians in appearance. They are slender with long tongues. Their hairs are short, dense and gray. The chitin is overwhelmingly dark, brown spots often appear on the second and third terga of the abdomen, and sometimes a leather-brown band is found on the abdomen. The color of hair of drones is gray to grayish brown.

The carniolan is the quietest and most gentle race of the three major races. One may leave the combs outside of the hive for a long time and not a single bee of a good strain will move away. Their rhythm of brood production is very steep. They overwinter with small colonies and small food consumption. Brood rearing starts with the first income of pollen and thereafter fast development occurs.

During summer the carniolan maintains a large brood nest only if the pollen and nectar supply is adequate. The brood rearing will be limited in cases of a poor or failing honey flow. After the fall honey flow, brood rearing will rapidly decrease, conserving stores for the long winter ahead. Overwintering is very good even under unfavorable climatic conditions.

Swarming with the present-day commercial strains, such as the Hastings Strain, is about the same as Italians. Carniolans have a very good sense of orientation, and no inclination to robbing. Their use of propolis is very small.

Diseases of the brood are almost unknown in the homeland of the Carniolan. This seems to be a particular feature of this race, because no special hygienic measures are practiced in the apiaries there. No explanation of this curious fact has been found.

The Carniolan developed in a part of Europe where the climate is influenced by strong continental air movements that result in long severe winters and hot summers following a short spring. Corresponding to these circumstances, the characteristic features of this bee are vitality and a fast, energetic reaction to any changes in environment. It is well adapted to regions with a heavy spring flow. The early honey flows and long, often severe winters in central Europe make the Carniolan the most popular race among beekeepers there. The Carniolan has become the second most widely distributed race throughout the world in recent years.

Now let us consider what the ideal bee for the Maine climate would look like. The first requirement would be its ability to overwinter in our climate, so that it would not be necessary to purchase expensive package bees from a southern bee breeder each spring. It also should be able to overwinter on a small amount of food so that the beekeeper can take more honey as surplus, thus increasing his honey yield. The bees should be able to build up rapidly in the spring to take advantage of our relatively early honey flow period, which begins approximately May 25th (in the Bangor area) from blueberry and apple, and finishes about July 10 with the clover flow. Also, it should possess a good sense of orientation, be quiet on the combs and gentle, and not inclined to swarm excessively. If the bee does not use propolis to excess, it would make hive manipulations much more pleasant.

After experiencing my share of winter losses and low average yields, I started thinking about a solution to this problem. In consulting other part-time and commercial beekeepers about this common problem, one common theme emerged, that of the correct choice of stock. After consulting the bee literature, I decided that the Carniolan would be the best choice for our climate.

In the spring of 1974 I purchased ten Carniolan queens and requeened some of my Italian colonies with them. They seemed to perform adequately as far as inclination to swarm and honey production were concerned, but would they winter? The spring of 1975 finally rolled around, and sure enough, the Carniolan colonies suffered only half the mortality of the Italian colonies. In the spring of 1975 I purchased enough Carniolan queens to requeen half of my colonies. Again, they seemed to perform satisfactorily and with the extra bonus of being extremely gentle.

In the spring of 1976, I carefully evaluated my apiaries which now consisted of 50% Carniolan, 25% Italian, and 25% Caucasian. The first check was the first of April and consisted of a quick check of the average number of frames the bees were covering. The Carniolans were covering an average of five frames of brood, and the Italians were covering an average of eight. This was the situation on April 1st, and I wondered if the Carniolans could build up rapidly enough to be ready for blueberry pollination.

On May 1st, the Carniolans value in our climate was dramatically illustrated. The relative number of frames of brood covered was now the reverse. The Italian colonies were now weaker. This is an excellent example of spring dwindling of Italian colonies in Maine. I believe the reason for this is that the Italians begin to raise brood in the late winter and consume all of their reserve pollen stores. The Carniolans, not beginning their brood rearing until pollen was available (about April 10th) could supplement their reserves with pollen from the field and maintain an uninterrupted cycle of brood rearing.

It must be emphasized here that no supplemental feeding of sugar syrup or pollen supplements was provided during the evaluation period. I have experimented with feeding pollen supplements and have abandoned the practice because of economic considerations and because of the Carniolan's ability to build up rapidly with pollen from the field, saving me both time and money.

The bee that is most likely to provide adequate returns to the beekeeper, be they hobbyist or commercial, will be the one requiring the least amount of care and expense. In my opinion, the Carniolan has this ability and I can foresee the day that she is the favored race in the state of Maine. A bit of advice to beekeepers experiencing heavy winter losses is to try the Carniolan before you give up in despair. You certainly would not be very successful raising orange trees in Maine. Why try to raise bees that are not climatically adapted?

The average yield of honey is increased by the use of the Carniolan because of her conservative use of the winter food supply. A typical average colony that I had setting on platform scales used 35 pounds of honey from the first of November 1975 to the first of April 1976. I maintain these bees in a one and one-half story hive for pollination rental, and this is the minimum size hive in which one should attempt to winter bees. Also, do not forget to provide top insulation, reduced lower entrance and some type of entrance in the upper hive body for ventilation purposes.

One other thing that might help the honey production would be to scatter the seeds of purple loosestrife and white sweet clover. Be sure to get sweet clover seed and not white dutch clover, as they are quite different. These seeds can be scattered along the roadside and other favorable areas.

RAISING BEES IN MAINE

To anyone with an appreciation for the tranquility that nature imparts, beekeeping is usually irresistible. Although a strong back is to a large degree indispensable, equally necessary are perseverance and attention to detail. The importance of learning from the experiences of others as well as familiarity with standard reference works cannot be overemphasized. Although beekeeping does not represent a major agricultural industry here, the beekeeper will find Maine to be a hospitable habitat.

The beginning beekeeper should purchase the basic books and

pamphlets that will serve as an introduction. The most practical source of information for the beginner is a pamphlet entitled *Starting Right With Bees.* Also of great value are Roger A. Morse, *The Complete Guide to Beekeeping;* Roger A. Morse, *Bees and Beekeeping;* Richard Taylor, *The Joys of Beekeeping;* and *The ABC & XYZ of Bee Culture.*

Here in Maine there is an enormous variation in suitable bee pasturage. The first thing one should do is to conduct a botanical survey of the area where the hive or hives are to be kept. Inasmuch as bees will travel from two to four miles for nectar bearing plants, the wise beekeeper will make a topographical survey map, pin-point the hives and then draw a series of circles around them at distances of one mile. Within the areas thus designated a careful survey must be conducted. To identify potential nectar sources two items are useful: H. B. Lovell, *Honey Plants Manual,* and Roger T. Peterson & Margaret McKenny, *A Field Guide to Wildflowers of Northeastern and North Central North America.* After checking off the honey plants native to your region study Peterson's *Guide* for the purpose of ready identification. Some beekeepers expand the nectar supply by planting various items such as Buckwheat, White Sweet Clover, Vetch, Huban Clover, Trefoil, Anise Hysop, Purple Loosestrife, and Fireweed. Unfortunately such planting practices are not consistently economical for the professional beekeeper. However, to the average beekeeper with a few hives it can be quite productive. In most Maine areas the spring, late-summer, and early-fall availability of honey plants is rather consistent. The heart of summer is frequently a difficult period for bee forage. Fruit bloom has passed and Golden Rod has yet to blossom. This is the season when the beekeeper must pay close attention to the cultivation of honey plants if the need dictates. Plants, shrubs and trees planted for decorative reasons should also be nectar bearing if at all possible or practical. A variety of such plants are obtainable from Pellett Gardens, Atlantic, Iowa 50022. A free catalogue is available upon request. Among the shrubs and trees planted for ornamental purposes, as well as bee forage, are White Alder, Bee Bee Tree, Basswood, Golden-Rain Tree, Honeysuckle, and Negundo Vitex, as well as fruit trees and berry bushes. A valuable guide to the trees of Maine is Lorin L. Dame & Henry Brooks, *Handbook of the Trees of New England With Ranges Throughout the United States and Canada.* Interestingly enough, some of the best bee pasturage will be found in cultivated, suburban areas. There the flora is often sufficiently varied to sustain a number of hives.

Now that you have carried out your botanical survey and stand convinced that a hive or hives will flourish, the next problem to be solved is the question of equipment. The most common and satisfactory hive used in Maine consists of a bottom board, two brood chambers containing ten frames each measuring 9 1/8 × 17 11/16th inches, a ventilated inner cover, and a metal top outer cover. The best hives have clear pine brood chambers, with dove-tailed corners for maximum strength; cyprus bottom boards; and clear pine covers. The ten frames in each brood chamber are also clear pine. A word of caution might be advanced at this point regarding "cut-rate" bee-hives. The problem is that the hives are frequently difficult to assemble due to carelessness in manufacturing, and the wood is often not properly dried or knot free. If knots are present they not only weaken frames but will dry up and work loose. The slight additional expense in purchasing top-grade equipment adds immeasurably to the durability of the hive.

To assemble the material necessary for one hive, the beginning beekeeper should purchase a standard hive, and an additional brood chamber with ten frames. It will also be necessary to purchase twenty sheets of wired foundation for the frames and forty support pins to hold the foundation rigid. The cost of the above will be approximately $45.85 plus sales tax and freight. Incidentally, it is essential that two brood chambers are utilized for each hive. One brood chamber will not permit the bees to accumulate the stores of honey that are necessary if they are to survive the frequently severe Maine winters.

Although beekeeping supply catalogues list a number of items that are a convenience or of debatable value to the beekeeper, there are certain basic tools that are quite necessary. Under the heading of indispensable are the hive tool and the smoker. The bees customarily glue the hive parts and frames firmly in place with a substance called propolis. Without the hive tool it would be almost impossible to open the hive or remove the frames. It will be discovered rather quickly that the judicious use of smoke is most helpful in distracting the bees during the periods of manipulation. The smoker will last almost as long as the hive tool if it is not abused. As to fuel for the smoker one will find through experience that rotten hard wood, thoroughly dried and broken into walnut sized fragments will produce a satisfactory smudge. When one adds short sections of corn cob and sumac spikes, also bone dry, the results are most adequate.

Beekeepers of long experience often affect a casual attitude toward bee stings. It is doubtful that one ever becomes completely oblivious to the pain a sting inflicts. It is foolish not to treat such eventualities with the respect they deserve. Care should always be exercised especially if one has yet to determine just how sensitive they are to bee venom. The use of a bee-veil and gloves is no occasion for embarrassment. Adequate protection not only makes beekeeping more enjoyable but is an exceedingly wise investment.

Occasionally second hand beekeeping equipment is available. It should be understood that the beginning beekeeper is taking an enormous risk in purchasing used bee hives. One of the gravest threats to profitable beekeeping is a disease called American Foul Brood. It is transmitted by a microscopic spore and there is no known cure. The only method of containment is to destory the infected bees and hive by burning them to ashes. Naturally, if you know the beekeeper and trust his expertise the risk can be minimized. Home-made hives should also be avoided. If they are poorly constructed with ill-fitting frames they are difficult to manipulate and for the most part cannot accommodate replacement parts.

Once the hive has been constructed some small matters must be attended to before the bees are introduced. One suggested alteration is the boring of a 5/8 inch hole in the hand grip on the front of the hive to improve the circulation of air. Climatic conditions here in Maine, both summer and winter, accelerate the accumulation of excessive amounts of moisture that endanger the survival of the bees. Also, some surfaces of the hive must be painted. All surfaces of the bottom board and the covers should be coated and the outsides and the edges of the two brood chambers should be painted, as well, with a good undercoat plus two coats of quality white exterior paint. White paint is commonly used to reflect the summer sun, because excessive heat is just as destructive as excessive moisture. It is well to remember that keeping the hive well painted enormously prolongs its life expectancy.

Just as important as the selection of quality equipment is the selection of the bees who will inhabit the hive. In the early years of American bee keeping the only bee available was the fractious "German Black." During the 1860's the black and gold banded Italian honeybee was introduced into the United States and soon became extraordinarily popular. Currently the Italian is the most common specie in Maine and has proved very well adapted to this climate. Although somewhat more aggressive than the Caucasian they are hard working and thrifty.

One common method of obtaining a swarm of bees, with a young queen, is to purchase a three pound package from one of the suppliers in the South. The packages are sent to the beekeeper by mail and delivery is, at the present time, a major problem in Maine. The U.S. Postal authorities no longer guarantee live delivery of bee packages. Some shippers have their own insurance arrangements, some insure the shipment through the Post Office. The experience of the past season was that approximately one package in three made it without major loss. The average price of a three pound package is approximately twenty dollars.

A convenient and much preferred method of obtaining bees is to purchase a swarm from a reputable Maine beekeeper. Three advantages are that live delivery is guaranteed, you get a larger quantity of bees and it's less expensive. Be warned, however, that a few beekeepers in Maine have taken advantage of the beginning beekeeper and have collected fees much in excess of package prices.

An area of very serious consequence to the beekeeper is that pertaining to bee diseases. Months or years of success and investment can be quite literally reduced to ashes. It is extremely important that the beginning beekeeper carefully study the symptoms associated with American Foul Brood, Bacillus larvae. As mentioned earlier nothing can be done to cure it once an outbreak occurs. As the stricken hive grows weaker other bees rob it of its honey store, thereby carrying the disease to their hives. In a remarkably short time an entire bee-yard or apiary can thus be overwhelmed.

Although no cure has been discovered there are methods available for the prevention of its development. The seasonal application of sulfathiazole and/or terramycin is currently very effective if correctly administered. The hive can be treated in early spring and late fall with no chance of contaminating the surplus honey. The most effective treatment is mixing with confectionary sugar and dusting the surface of the brood comb with it, or by feeding the colony sugar syrup with the drug dissolved therein. The former method is by far the easiest. All beekeeping supply dealers keep the drugs in stock. This is one aspect of beekeeping that must not be neglected, inasmuch as infection is an ever-present danger in Maine. For some reason American Foul Brood is especially virulent throughout the state. This is due in part to carelessness and ignorance among some uninformed beekeepers. The disease would also be more adequately contained if the Maine Department of Agriculture had sufficient funds available to carry out a full time program of inspection. A pamphlet that should be a part of every beekeeper's reference collection is *Diseases of Bees and Their Control*, Circular No. 527, published by Pa. State University, College of Agriculture, University Park, Pa.

As of this moment, Maine does have at least two part-time Bee Inspectors who are authorized by law to periodically examine all hives of bees and who can order the destruction of those infected with American Foul Brood. Unfortunately it is impossible for them to carry on an adequate program. They will, however, respond to a request for inspection within the limits of the time they have to spare. Contact can be made through the Division of Plant Industry, Maine Department of Agriculture, Augusta, Maine. The department also carries on a licensing program whereby all beekeepers are required by state law to register the number and location of their hives. This program of registration involves the payment of a minimum fee of two dollars, or ten cents per hive beyond a count of twenty. Registration is not an infringement upon the privacy of the beekeeper. The basic intent is to facilitate an effective program of inspection should funds ever become available. The fees collected are not enough to carry inspection forward beyond the casual methods now employed.

Another problem that must be touched upon involves the formidable impact of pesticides and insecticides on bee population. Here the diplomatic talents of the beekeeper will find full employment. We are all aware that over-kill is quite often the name of the game when it comes to using these poisons. This is especially true of the home gardener and regrettably, all too frequently, the case with the farmer. The injudicious use of pesticides is one of the largest causes of damage to the beekeeper. Every year tens of thousands of hives are destroyed throughout the nation. Here in Maine the problem is by no means minor. Farmers who spray fruit trees, berry bushes, and vegetables during bloom inflict staggering losses on the bees. Probably the most widely used pesticide ingredient, at the present time, is Sevin. This extremely poisonous substance is sufficiently slow acting to allow the bee to return to the hive with Sevin laden pollen. In due course the hive is so badly depopulated that survival is unlikely. One common source of Sevin poisoning is corn pollen. The corn farmers usually spray heavily to eliminate borers, and when the tassels are pollen laden the poisoning of the bee is inevitable. The only alternative is to try to educate the pesticide users as to other, equally effective, substances. With regard to plants in bloom, an attempt should be made to withhold pesticides until the bloom is past. Inasmuch as effective legislation banning indiscriminate, profligate use of pesticides does not exist, one's powers of intelligent persuasion must be fully used. It might be helpful to sweeten the argument by offering the farmer or gardener a few pounds of honey for a positive response.

Not to be overlooked is the picking of a site for your bee-yard or apiary. If you have space enough, choose a spot protected by a screen of trees or bushes from the northern winter winds. Face the hives south or southwest to take advantage of the warmth of the sun. Some shade should be available for the summer months. It is a definite advantage to place the hives on a platform raised about a foot from the ground. Not only does this protect the hives from dampness but also discourages ants, mice, and skunks from invading the hive. It also serves to keep tall grass from clogging the entrance of the hive. Incidentally, it would also be advantageous to place the hives so that they are as inconspicuous as possible. One current problem among beekeepers is theft. With beekeeping a rapidly growing avocation, thousands of too easily accessible hives are being stolen. This can be somewhat discouraged by burning one's name or initials deeply into the exterior of the brood chambers. In addition, vandalism has assumed disturbing proportions. Nitwits with hunting rifles or shotguns frequently take great delight in blasting away at beehives. The damage can well be imagined. Not to be overlooked are unfeeling juveniles with rocks in their hands. Finally, low visibility may be of benefit in dealing with your neighbors, at least until they grow accustomed to honeybees working their flowers. It is well to bear in mind that to the ill-informed, every stinging flying insect is a bee and will obviously be from your hive. Once again it will be politic to sweeten your apprehensive neighbors with a few pounds of honey. The last thing one needs is a public nuisance flap that necessitates the removal of the hives. Perhaps if your neighbor is a gardener or farmer, a conversational crash course in the value of honeybees as pollinators would be very much in order.

It is too bad that Maine does not have an active, progressive state association of beekeepers. The possibilities for dissemination of information would be great indeed. Few things are of greater value to the beginning beekeeper than the advice and assistance of his experienced peers. Lacking such an organization, the next best thing to do is to make contact with beekeepers in your area.

Obtain a list of registered beekeepers prepared for free distribution by the Maine Department of Agriculture. Arranged alphabetically by county and town, it should prove helpful in identifying beekeepers in one's section of the state. The list also gives the numbers of hives per beekeeper.

A question that is asked with increasing frequency is whether or not one could make a living as a beekeeper in Maine. First of all one should define what they mean by "a living." The number of fulltime beekeepers who derive their entire income from beekeeping in Maine can be counted on one hand. However, the people who supplement their income through beekeeping are indeed quite numerous. The average yield per hive in Maine is approximately fifty to seventy-five pounds. This is a rather low yield compared to many other sections of the country. Under the circumstances, the number of hives necessary for a survival income could not be less than several hundred.

At the present time there are a great many beekeepers who are gradually increasing the number of their hives in anticipation of a retirement income. In addition, they are adding labor-saving devices as well as honey processing equipment, on an annual basis, in preparation for this eventuality. In a remarkably short period of time the sale of honey will defray the cost of the original investment and will pay for gradual expansion. The best plan is to begin very modestly, preferably with one hive, and develop expertise through experience. Many active beekeepers who are familiar with and practice modern methods, as well as most dealers in beekeeping equipment, are happy to answer questions and provide practical advice. The beginning beekeeper must expect to pay out approximately $110.00 as an initial investment. This includes two shallow supers in which the bees store the surplus honey harvested by the beekeeper.

If, at this point, you are inclined to proceed, some serious consideration must be given to the problem of processing the honey. The only practical approach for the individual who plans to harvest large quantities of honey is to invest in an extractor and an electric uncapping knife—both expensive items. Without them, the honey cannot be removed without destroying the comb. Naturally, if one is not seeking liquid honey in bulk, this equipment is not essential.

Finally, it must be observed that the longer one functions as a beekeeper the more convinced he becomes that the learning process will never end. One soon discovers that it is absolutely impossible to consistently predict the yield of any specific hive. The same holds true in establishing a routine behavior pattern for the bees. Truly this communal creature moves in many and mysterious ways. The use of common sense, bolstered by constant study, is the only path to any degree of success. But that success can bring joy.

Chapter 2

Farmstead Fowl

The poultry industry is one of the largest industries in Maine, as anyone can readily tell just from the number of windowless, metal-sheathed buildings that clutter the landscape everywhere. Along with an idea of the size and distribution of the industry, these buildings should also give you an idea of what chicken farming in the industrial age has become; i.e., the farmers are no longer farmers, but businessmen, promoters, and contract growers. The chickens are still chickens, but with a multitude of changes in their diet, genetics, and living conditions. And as in almost any large industry there is waste—in the name of economy—which can directly benefit the small Maine homesteader.

STALKING THE LOW-COST CHICKEN

My wife and I consider ourselves vegetarians. We do not use any animal products which necessitate the killing or abuse of animals in their procurement. We therefore were receptive to the idea of securing some hens for laying purposes which should allow us to bypass hatcheries or other farmers, since we didn't want to buy pullets only, and thus condemn an equal number of hatched cockerels to the soup pot. And we were equally interested in sharing a little bit of the good life with some birds which had already been mistreated in the name of profit. Nor did we object to saving money. With this in mind, let me tell you how we began our flock of excellent layers with no capital outlay except for feed and housing—the chickens were free, of superior laying strains, and almost ready to lay.

The average battery hen or cage-layer lives in a temperature-controlled house along with some ten to fifty thousand other individuals. She shares a small, wire cage with one or two cage-mates in such close quarters that she cannot fully extend even one wing at a time. She cannot scratch for her food, dust bathe, or exercise. If she is at the bottom of the pecking order she cannot escape her tormenters. If she's sick, she is ignored, and if she dies (either from disease, or as many do in the first few weeks of confinement, from "stress"), she becomes a rotting carpet for her remaining fellows until her carcass is discovered and removed. Her toes curl unnaturally from the wire floor, and her body may be caked with manure, since she has never known rain. She is kept under an extended artificial day, and boy, does she ever lay eggs . . . for a while. After a year to eighteen months she is worn out, used up, pale-skinned, skinny, in poor feather, and fit only to be dragged from her cage for one last ride—to be turned into soup. In short, she is no longer an economical asset to her owners, one of the large growers who pay the farmer a few cents a day to feed and house her and collect her eggs.

She is now ready for you. Approach your neighborhood egg producer when a flock is about ready to leave, or has just left his farm, and you may be able to get some of these battered beasts for yourself. Seldom do the pickup crews get all the birds. Many will probably have escaped in the process of being transferred from cage to truck, and they will gradually creep out of the bushes in the next few days. Some, perhaps many, will be found in the cellar, where they have been living for months after escaping from their cages, as quite a few do. These latter birds will be dirty and covered with manure, but don't overlook them because they are the better layers of the two groups. The farmer is likely to say— "catch 'em and they're yours", because he is interested in getting his facilities cleaned up and ready for the next flock, and these old, worn out birds are of no use to him. Or he might let you have them in exchange for a little work in helping him clean his barn. Don't give him cash for them because at that stage they aren't worth cash—you haven't yet rejuvenated them.

The dung-covered birds you found in the poultry barn basement should be cleaned up with a bucket of warm water on a warm day in the summer, or just left to clean themselves if the weather is bad. Dust them with Sevin dust. No, it isn't organic, but it is the only sure way to get rid of mites and lice, and after two treatments at ten day intervals, you can go back to being organic again, and your flock will be rid of the annoyance and production loss of parasites. Within about three weeks to a month most if not all of these birds should be laying eggs. Since they have been living in the basement under conditions of very dim light, although they are over a year into their laying life they will have laid very few eggs, and their egg machinery is just waiting for your good food, clean water, and good care. They should lay well for several more years, since their genetic lineage is from the best laying stock of their breed that the hatcheries could produce.

The creepy-looking bald things from the cages will take a lot longer to begin laying, since they have truly been used up. A couple of months or more will pass before they do much of anything for you, and even then they will never again be the high-producing birds they were in their youth. We have given a few of these birds to an old rooster and he seems content with them, and they are gradually getting used to using their feet and wings again and doing good natural chicken things.

Now, you ask, why you should bother with these old things instead of new chicks or ready-to-lay pullets? My answer is that if you are only concerned with eggs and meat you probably shouldn't. But an ever increasing number of homesteaders are vegetarian or at least have some compassion for these mistreated animals. (Note: I am not speaking in some wishy-washy anthropomorphic sense here. I do not worry that these chickens have been deprived of their "rights." I do suggest that they are, many of them, in physical pain through lack of attention or care, and that they exist in confinement only to secure profit for a corporation until such time as they have their throats slit to make one last dollar). Nor should you overlook some economic advantages of your own; they are adult birds, and with proper care should lay soon; they are from families of proven high producing stock—they have been vaccinated against one or more common diseases, and, as mentioned, they are cheap or free.

It is probably best not to be too hasty in mixing these animals with birds from another source, since they may have been vaccinated against a disease of which, although they will not become sick, they can still be carriers. At any rate it is generally good practice to keep your birds in several small flocks rather than one large one, so you can still have them from more than one source. This way any disease problems you encounter will likely be the result of dirty housing or yards, failure to isolate sick birds or bury dead ones carefully, or something introduced by another person or animal.

Don't think about this source of birds if you are impatient, unwilling to spend some time helping your new flock get started, or if you really aren't concerned about humane treatment of animals.

You will not get rich from your egg production, but you shouldn't go broke, either. We started with three hens and a rooster in the summer of 1973. Now, at the end of January, 1975, we have fifteen hens, two roosters, two ducks, and two geese. Our eggs during this time have averaged about 90 to 95 cents per dozen, have been almost exclusively large to jumbo, and have included the cost of feed for all those other birds in addition to the laying hens! You can see that if we were concerned with profit only and had not acquired the ducks, geese, and roosters, our eggs would have been a bargain indeed. But even with all those non-producing hangers-on, we're breaking even compared to store prices, and have the advantage of fresh eggs from our own friendly flock.

Further, we intend to drastically reduce our cost of grain this summer by growing our own field corn and drying it. Since corn can constitute up to fifty percent of the ration (particularly during the winter months when a little extra fat is good for the birds), and since they will also eat mangel beets, squash, and table scraps in abundance, the small homesteader can cut his grain costs by as big a percentage as he has ambition.

And when they stop laying? Well, we think even old chickens deserve a retirement, although they certainly don't need the high protein ration of laying birds. So our old hens don't go under the axe. They will go into a large pen, a range where they can forage for much of their food supply, which we will supplement with the aforementioned homegrown grains, and garden produce. This is not an idle dream but a proven fact, since several birds wintered over a Maine winter after escaping from a neighbor's flock with only minimal shelter and what food they could scrounge from bird-feeders and the woods. They may have struggled a bit, but they were fat and well-feathered in the spring.

Don't be surprised if, even in the face of a year or two of successful egg production and thriving hens, your farmer neighbors laugh at your efforts. They have been educated, unfortunately, by agribusiness, trade publications, and the bigger is better and newest is best philosophy so peculiar to our country and our times. They will scoff at your idea that a chicken needs more than food, water, and light, and few but the most strict vegetarians will admit that an egg from a happy, comfortable hen is preferable to one forced from a caged bird. But then, you aren't homesteading to be admired or popular, and if your chickens fit in well with your lifestyle you have no one else to answer to. And my own experience has shown me that even the state Extension Service often fails to understand your problems. When I wrote to the poultry office at Orono recently seeking some general information I admitted that a large part of my birds' diet was composed of squash, mangels, and table scraps. The experts were quick to point out that this diet was "too high in fiber" and would lower my egg production. This, they no doubt knew to be true in every case, though my own practical experience was that my egg production was better than ever, even though my chickens were at that time two years past their prime.

We are not alone, either in our desire to see our livestock treated humanely, or in our experiment with cage hens, since friends of ours have been successful in following our lead. Information on care, feeding, and housing is available from your county extension agent.

CAN YOU AFFORD FREE CHICK?

Many homesteaders, part-time farmers, and others with at least one foot planted firmly in the soil have taken their first taste of self-sufficiency by raising chickens. A flock of poultry wandering around the dooryard clucking contentedly does much to turn a house into a homestead, and everyone who has raised chickens sings the praises of home grown, home cooked chickens over the poly-wrapped supermarket variety.

It is a predictable sign of spring to see feed stores advertising free chicks, often along with coffee and donuts, and this has become a popular method of starting a backyard flock without laying out a lot of cash. Almost everyone has heard of the chicken's relatively efficient conversion of grain to meat, and after all, if it's free, how much can it cost you?

It can, unfortunately, cost you plenty. Basically, there are two types of chickens—layers and broilers (growers). It is the broiler that converts feed efficiently and reaches a hefty four pounds in only eight- or nine weeks. The layer will consume about 20 pounds of grain during its first 20 weeks, at which time it will start to repay your investment by producing eggs . . . if it happens to be a she. The problem is that the free chicks offered to aspiring poultry farmers are neither hens nor broilers; they are laying breed cockerels and can neither lay eggs nor convert feed efficiently. When eggs are hatched out to raise replacement hens, Mother Nature still has enough of a hand in the matter to decide that about half those eggs will hatch out females and the other half males. The females become replacement pullets, the males are cockerels and are of no further use. No further use, that is, unless a feed company will take these chicks off the hatchery's hands and give them away to folks who happily pour 20 pounds of expensive mash into them, all the while thinking they'll soon be eating almost free chicken.

Let's take a look at the economics of the situation. On the one hand, take 10 free chicks and raise them to about four pounds each, which will take 20 weeks and about 20 pounds of feed per bird, or 200 pounds for the flock. With grower mash at about $8 per 100 pound bag you'll have spent $16 for about 40 pounds of chicken. Bear in mind that this is liveweight (i.e., feathers, blood-'n-guts, and all) and the dressed out birds will weigh in at about 70 percent of the liveweight. On a liveweight basis, then, your chicken costs you about 40 cents per pound when raised from free chicks (or 57 cents a pound for 28 pounds dressed out.)

The alternative is to buy broiler chicks. These won't be free, but they should cost you no more than 50 cents each. Suppose you buy 11 of these and raise them to four pounds. You'll need eight or nine weeks and 100 pounds of grower mash (about 2 1/4 pounds of grain per pound of chicken) to wind up with 44 pounds of chicken (liveweight). That 44 pounds costs you the $5.50 (11 × .50) for chicks and $8 for 100 pounds of feed, a total of $13.50 or about 30 cents per pound.

What it boils down to is that if you get as many as 50 "free" chicks it's costing you $20 more to raise those birds than if you paid 50 cents per chick for broilers. And while it's true that homesteaders and part-time farmers aren't just in it for the money, most of us can ill afford to throw the stuff away.

One word of caution before you rush out in the spring to fill your brooder with purchased broilers. People (including those who sell chicks) have a tendency to label anything that doesn't lay eggs a broiler, so when you're buying broiler chicks be sure they actually are birds of a breed intended to be raised for meat, and not merely the males of a laying breed.

TAKE ANTIQUE CHICKENS OFF THE SHELF

The collecting of antiquities is currently in vogue all over the world. One can visit antique shops in almost any city, town, or hamlet throughout the United States. Certainly there is little novelty in the formation of clubs dedicated to the collecting and preservation of antiquities. What may be surprising to many, however, is the existence in the United States of an organization dedicated to the preservation of rare breeds of poultry called the Society for the Preservation of Poultry Antiquities.

At a time when megalopolis and urban sprawl have become a part of our vocabulary, millions of people never see live poultry. The country fairs are certainly still flourishing in Maine; in fact, they seem to be increasing after suffering a decline in popularity. There are exhibitors of cattle, sheep, and sundry other exhibits, but no poultry. One by one the fairs here in Maine have discontinued poultry exhibits even though there are enough poultry fanciers in Maine alone to fill an exhibition hall with exotic and rare breeds such as would astound most viewers. For the first time most would realize that a "chicken is not just a chicken!" A number of the New Hampshire fairs have excellent poultry shows and the exhibition of poultry is on the increase in that state. There were actually a few years in the late 1960's and early 1970's when there was not a single poultry show in Maine. Thanks to a small group of dedicated poultry enthusiasts such as C.R. Woodman of Readfield, Maine, the Northern New England Bird Fanciers' Association has been formed and once again Maine has its own poultry show. Perhaps some of the readers of this article had an opportunity to view the 1976 show held at the Lewiston Shopping Center where exhibitors from several states competed for various awards. Many varieties of standard fowl, bantams, pigeons, turkeys, ducks, geese, and guinea fowl were displayed for the thousands of shoppers to view, many of whom exclaimed with utter astonishment at some of the rare and equisite breeds which formed a kaleidoscope of colors with their lovely plummage.

Although the term poultry covers everything from geese to guinea fowl, the remainder of this article will focus upon chickens, large fowl or standards as opposed to bantams. Since the poultry industry is now conducted primarily in the form of large corporations, a constant effort is being made to produce so-called sex-linked birds that will shell out eggs at a rate that would shock the farmer of yesterday who earned a good living from keeping a few hundred Rhode Island Reds, White Plymouth Rocks, Barred Rocks, or White Leghorns to name some of the more popular breeds of twenty-five to fifty years ago. It used to be that when hens ceased to lay and began to molt or shed their feathers, one could be assured of delicious chicken dinners if he did not wish to keep his laying hens over until such time as they would begin laying again. Today hens bred for egg production are not worth dressing off. Every ounce of energy and goodness in the bird has been expended in the production of eggs; there is nothing left worth salvaging when the hens stop laying. This is fine; this is big business. However, one might ponder upon the origin of these sex-linked birds. From where do the blood lines originate? All commercial breeds today were developed from older breeds of poultry, most of which are listed and described in a book called the *American Standard of Perfection*. It is the concern of poultry fanciers and poultry geneticists in the United States, Canada, and other nations in the world that unless efforts are made to preserve the old breeds, there will come a time when it will be impossible for geneticists at agricultural research laboratories to develop and improve upon the commercial strains, a necessity if the commercial poultry industry is to remain alive. There are, of course, less pragmatic reasons for the concern about the preservation of antique breeds of poultry. These reasons are similar to those which have generated herculean efforts on the part of conservationists to protect and perpetuate the whooping crane, the silver-tipped grizzly, the eagle, or the redwood. It is deplorable and disconcerting to many whenever any species becomes extinct.

There is a long list of standard breeds that are on the endangered list, and each year the list seems to grow longer. Some of these breeds are seldom seen even in the largest poultry shows in the United States and Canada. How many have ever heard of, let alone seen, such breeds as the Dominique, La Fleche, Dorkings, Buff Catalans and Whitefaced Black Spanish to name but just a few? As the years pass, the endangered list becomes longer and longer. The Mottled Houdan, Speckled Sussex, Lamona, and even the New Hampshire Red have been added to the more recent list. It is not just the members of the Society for the Preservation of Poultry Antiquities who are striving to keep these breeds from dying out. There are hundreds of other clubs and thousands of exhibitors and nonexhibitors, who belong to no special organization, who are helping to preserve our old and lovely breeds. The problem is that some breeds such as the La Fleche have become so scarce that it is difficult for one to acquire the breed and even more difficult to add new blood lines to one's breeding stock. One of the services that the Society for the Preservation of Poultry Antiquities provides is a directory of all its members and the breeds and variety of breeds which each has. However, even when one locates a breeder of one of the rare breeds for which he has been searching, the breeder may not have extras to spare. If he does, the cost of shipping live birds is outrageously costly and the purchaser must bear the cost which very often is much higher than the cost of the breeding stock. Few breeders will ship chicks as they do not usually hatch on a large scale. Many will ship eggs, but it is always a gamble that they will arrive in good order and the hatchability of shipped eggs is very often low or nil.

No one who has never attended a large poultry show and who is unfamiliar with all but a few more common breeds, it may come as startling news that there are around 120 different varieties and breeds that have been admitted to the *Standard of Perfection*. There are many other breeds and varieties that hopefully can be preserved and admitted in the future. To be accepted to the *Standard of Perfection*, a breed or variety of that breed must be identifiable by certain characteristics such as color pattern, type of comb, size, and shape, all breeds have been divided into classes. There are at present nine classes: the American, English, French, Mediterranean, Asian, Hamburg, Continental, Polish, and Old English and Modern Games. Almost every breed has more than one variety. The familiar White Leghorn is but one of twelve varieties belonging to the Leghorn breed. To cite a few examples of what constitutes variety, there are Single-Comb Dark Brown Leghorns, Rose-Comb Dark Brown Leghorns, and Single-Comb Black, Buff, Silver, Red, Black-Tailed Red, and Columbian Leghorns. There are fowl with single combs, rose combs, forked combs, pea combs; there are breeds with crests, beards, muffs, and feathered legs; there are those with five toes; there is one breed with white faces; the Auraucanas lay blue, pink and green eggs (the low-cholesterol content often attributed to their eggs is fallacious); and, of course, there are breeds and varieties with every color pattern imaginable, including blue. The many color patterns and varieties of combs make raising fancy poultry challenging and exciting. Poultry enthusiasts can only be stereotyped in as much as they share a common interest in raising poultry for aesthetic reasons. Doctors, lawyers, teachers, men and women from all occupations, youth and the aged can all be found at a major poultry show sharing their experiences, their successes, and failures. The winners are usually magnanimous in offering helpful tips to the losers, and a convivial atmosphere intermingles with a cacophony of sounds characteristic of the poultry world.

One of the goals that most poultry hobbyists have is to encourage others to take up the hobby. The more poultry enthusiasts there are over the country, the better the chances our rare breeds will have to survive and to perpetuate themselves; the more people involved with poultry, the more clubs and consequently more shows will spring up over the country. It is a healthy and wholesome hobby. All shows attempt to offer incentives to young people to exhibit birds. I am in the field of education, and each year I realize great satisfaction from taking some of my own birds to my high school and giving lectures on poultry genetics to the biology classes. Having live birds in the biology lecture room always generates a great interest among the students. For most, it is the first time that they have viewed anything but the more common breeds, and they are amazed at the variations of traits in different breeds. A show and tell lecture with poultry is a guarantee of a successful day in the classroom. Like any other hobby, raising a small flock of poultry is therapeutic. Perhaps it is more so than many hobbies because it is more demanding. Poultry should be fed and watered twice a day. Usually when I return to my home in the afternoon, I am carrying an attache case filled with work that must be completed before the following day. Having to care for the poultry gives me a respite from mental pressures. I enjoy feeding and watering the birds, gathering the eggs, checking out the feathering of one of my choice birds, or working with them so that they will not become nervous when they are handled by judges during a show. My birds have compelled me to take that break from my regular work; I feel more relaxed and ready to cope with the evening's work.

For some poultry hobbyists, tracing the origin of their breed or breeds can be fascinating. The classes of most breeds denote the area of their origin. However, tracing the origin of breeds of poultry can be as complicated as attempting to trace the origin of races, providing, of course, one recognizes that there are distinct races. To be more succinct, Orpingtons are a breed that belongs to the English class since some of its varieties were first developed in England. However, the Orpington was developed in part from Asiatic and Mediterranean breeds. Since facets of my work involve rather extensive travel, I frequently have the opportunity to observe small flocks of poultry running about in the streets and yards of villages around the world. What I am always looking for but seldom see is a flock of one of our standard breeds. Usually the flocks are typical barnyard flocks that have been crossed and recrossed. Surprising, perhaps, are the obvious traits of our American breeds that frequently predominate in the flocks I observe. I specialize in English and French breeds. Along the Champs Elysees, the most fashionable promenade in Paris, there is a section where there are a number of pet shops clustered together. In addition to vampire bats and assorted snakes, there are the more mundane pets including cages of poultry. Each time I pass the shops, I look in vain for Crevecoeurs, Houdans, La Fleche, or Faverolles. Instead I see Rhode Island Reds, White Rocks, Leghorns, and Asiatic breeds. The French breeds are named after the towns or villages where they were first developed. For example, the Mottled Houdan (one of my favorites) was, according to the records, developed in Houdan, a town of 2,100 people located about forty miles from Paris. I once motored out there to see the place of origin of one of my breeds. The fertile green fields and the white-washed houses are still there, but the Houdan has all but vanished from the place of its origin. There are times, however, when my vigil and perseverance are rewarded. For example, one day while hiking along the country roads in Flanders, Belgium, I observed the finest flock of Salmon Faverolles I have ever seen. Although it was not in France that I observed them, the village and setting were somewhat similar to Faverolle, the place of their origin.

One thing I have observed from traveling and breeding poultry is that there seems to be a remarkable likeness between the temperment of the breeds and the indigenous population where the breeds originated. For example, Mediterranean breeds are usually more excitable than English breeds, and the Asiatic breeds such as the Brahmas and Cochins maintain, with few exceptions, a perpetual Asiatic calm.

Few poultry enthusiasts realize a profit from their hobby unless they are able to sell many of their superfluous stock at a good price, but raising poultry does offer some monetary returns directly or indirectly. Many rare breeds are good layers, and most originally were bred for utilitarian purposes. This means that they are useful for both meat and eggs. One can enjoy his own fresh eggs, sell the surplus to help pay for the grain, and savor fresh chicken when it comes time to cull out the flock. Frequently there are opportunities for one to sell hatching eggs, baby chicks, and breeding stock within the area or across the nation if one wishes to advertise, especially in *The Poultry Press*, the only publication completely devoted to the needs and interests of the poultry fancier and the preservation of all breeds of poultry. If one takes pride in possessing a beautiful flock, he should also take pride in keeping his pens clean and the floor covered with fresh shavings or some other kind of clean litter. Usually if one keeps poultry, he also maintains a garden. There is no better fertilizer available than poultry manure, providing it is used with some discretion, to make vegetable and flower gardens, lawns, and fields productive. If one cleans out his pens frequently, there will be little odor from the soiled litter; yet, it will contain enough nitrogen to induce almost anything to grow more luxuriantly.

Finally, for those who are contemplating raising a small flock, the following tidbits of information may prove helpful. The selection of a breed or breeds is always a problem. There are a few hatcheries in the United States that specialize in hatching rare breeds. Murray McMurray of Webster City, Iowa, advertises sixty-seven varieties. They publish a beautiful and colorful catalogue, free upon request, which is extremely helpful to the novice as almost every breed and variety is pictured, most of them in color. There are rare breed specials where one may acquire a rainbow of chicks at very moderate prices. This is tempting and is a quick way of sampling a number of breeds. It can be exciting as chicks change in appearance from day to day, and soon one finds himself with a mosaic of rapidly maturing pullets and cockerels. In many cases one could end up with perhaps only one of certain breeds. For those who are more serious about becoming successful as exhibitors it is better to focus upon as few breeds as possible. If one is confronted with a problem of lack of space, it is more feasible to limit himself to one or two breeds. The more birds of a single variety, the better the odds that there will be a few which will somewhat approach the approximate standards set for that particular breed or variety. If one wishes to hatch his own chickens from his own breeding stock, it is advisable to have more than a single pen of breeders. One must consider the mortality rate even among healthy flocks. There is always an element of risk in showing birds because they are vulnerable to a number of diseases, especially forms of respiratory diseases. It is often tempting to purchase a large number of chicks, especially at bargain prices. Remember that the price of grain is very high, and growing chicks have prodigious appetites.

One may soon lament the fact that he yielded to the temptation to purchase too many. Twenty-five to fifty chicks is what I personally recommend. As far as breeds are concerned, that is a matter upon which only the individual can decide. It may take several years of experimenting before one can settle upon one or more breeds which he really feels are his favorites. Hopefully raising and perhaps exhibiting poultry will become a family hobby, not just a hobby for an individual member of the family. There is a dire need for more family oriented hobbies at a time when families seem to find little time to share experiences. Over the years that I have been attending shows throughout New England, I have observed that some of the most successful exhibitors are those who make up a family team. Last of all, if one wishes to exhibit or just learn more about raising poultry, he should join a poultry association. It is only through the strength of hundreds of clubs and associations throughout the United States and other nations of the world that many of the rare and exotic breeds will be saved from extinction.

IN PRAISE OF CHICKENS

Chickens. Just say the word, and for most people adjectives such as dumb, dirty and smelly come to mind. Nearly everyone has a cousin or an aunt who kept chickens at one time or another, and the chief impressions left by a summer vacation at their house are likely to be a hoe blade chopping into rock-solid chicken droppings, the eye-stinging odor of ammonia, and grandma getting fresh manure stuck to the bottom of her bedroom slippers when she went out to collect the early morning eggs.

Friend, it doesn't have to be that way. Along with spiders and snakes, chickens are among the most maligned animals on earth. The potential of these versatile birds is grossly underestimated. With a healthy dose of common sense, keeping them can be little more trouble (and considerably less work) than maintaining the family lawnmower.

And the rewards are great. Not only will you replace your pale, aged, store-bought eggs with a fresh, yellow-yolked product that stays together in the pan and tastes like the real thing, you will also gain some free labor for the homestead. If allowed to run outside, chickens will improve your land.

Living out-of-doors is the key to easy chicken maintenance, low chicken-keeping costs, chicken happiness and chicken health. Good milk comes from contented cows, the advertisement used to say, and you haven't seen a chicken (or tasted an egg) until you've seen a chicken in it natural surroundings.

Out on the land, chickens will thrive on any rough ground, provided only that it is not too wet, pulling weeds, scratching the soil and eating grubs and insects of all sorts. In pasture, they will scrape out matted grass during their foraging and eat weed seeds, fertilizing the ground all the while. Let into the garden after harvest, they will scrounge for spilled corn and other remnants, pick at weeds, gobble up insect pests, and add nitrogen-rich fertilizer to the soil.

Although you would never know it from reading the label on the standard chicken feed bag, chickens are omnivores; that is, they will eat nearly everything that qualifies as food. And in tribute to their innate good sense, they will balance their own diet if given the chance. On the range they will eat any but a very few varieties of green things, supplemented by worms, grubs and anything else they can catch. The diet is strikingly similar to that of wild birds, of course, and theoretically it could maintain the chicken well. However, sustained egg production requires a supplementary protein feed.

Forget buying antibiotic and hormone-laden, highly charged chicken feed. Foraging birds can be well maintained on "scratch feed," an assortment of corn, oats and other cracked grains. Additional protein can be added by feeding a bit of protein mash, or supplementing the chicken's fare with table scraps, including fish and meat leftovers, and soured milk poured on their grain.

Liberated chickens get all the trace minerals they need from forage rather than from plastic bags, except perhaps calcium. Large amounts of calcium go into eggshell production, and this must be replaced by allowing the birds access to soured milk, clam or oyster shells, or their own eggshells. Eggshells should be crushed and dried in a low oven before being fed to the chickens, however, lest the birds develop a taste for the egg residue that remains on them and on a day of desperation, begin feeding on their own freshly laid eggs.

Constant access to water is probably more important to egg production than large amounts of protein supplement. If water is not available during the critical egg-formation hours, production may drop as much as 50 percent. Chickens will naturally seek out the sandy grit they need for digestion and dirt for anti-lice dust baths, if given the chance.

All this is not to say that chickens shouldn't be restrained in some way. A five-foot high chicken wire fence with one-inch mesh and a tight-fitting gate is a good investment, not only to keep the chickens confined when you want them to stay put (to prevent them from laying in the woods, for example), but also to keep predators away. Mount the fence on solid posts in a convenient spot near the barn or garden, and bury its bottom edge at least six inches deep to discourage raccoons, skunks, mink, coyotes, and other wild beasts that relish a meal of fresh chicken or eggs. The amount of space you enclose should be governed by the number of chickens you are planning to keep and your own common sense. Look at the area you are considering for the chicken yard. Would you be comfortable in there if you were a chicken?

HOMEGROWN CHICKENS

My husband says store-bought chicken tastes like a cross between Zonite and Lysol. I know I don't like to cook chicken and have the bones turn black. So every year I raise 25 cockerel chicks to put in the freezer. With good luck, I'm able to raise at least 23 of these. But good luck is mostly good care.

I start with good, clean chickens from a reputable hatchery. The Cornish Cross chicks are best for eating, as they have broad breasts and extra white meat.

In addition to the 25 Cornish Cross cockerels for the freezer, I also start 25 White Leghorn pullet chicks for egg layers. They can be started together and separated when they reach four weeks of age.

Order the chicks at least three weeks ahead of the date you want them. I start them the last week in May. At this time of year, you can brood them safely with a light bulb. In a tight little coop a 150-watt bulb will keep them warm enough; in other places you may need a 250-watt heat bulb. Never use a heat generating bulb in an ordinary light socket. Be sure it is a porcelain socket so you won't risk fire.

I fence off a corner of the old milk house with wire screens. Discarded whole window screens are good. You must have a rodent and predator tight place for the chicks. It must be water tight and draft free.

A big cardboard box (the type an appliance would come in) placed in a corner of your basement or laundry room can be used for the first week if necessary. Line with layers of newspaper and sprinkle with dry sand. Keep it clean; change papers daily. This is safe only if you have no cats or rats in the area.

Suspend the light bulb about a foot off the floor over the chicks. Sprinkle fine chick scratch on the newspaper—right on the floor at first. Do this a little at a time and often. Do not use chick starter mash until chicks are at least four days old or you will have trouble with constipation. Fine ground corn (the kind you feed birds) can be used if you can't obtain scratch. Buy an inexpensive chick waterer as you have to have a water container that the chicks can't get in and get wet. It's best to buy two waterers and give the chicks water in one and milk in the other. Milk protects the chicks against "coccidiosis" and makes them grow faster. Meat from milk fed chickens seems to be juicier, more tender and flavorful.

You can tell if the chicks are comfortable by their behavior. If they spread out like a carpet under the light they are comfortable. If they crowd into the farthest corner and especially if their bills are open, they are too hot. Or if they pile up under the light they are too cold. Adjust the size of your bulb and height off the floor until they act comfortable. Do not get the bulb too near the floor, because of the danger of fire.

Chickens should never be crowded or in damp conditions. Avoid chilling and overheating. Natural enemies of chicks are cats, rats, snakes, weasels, hawks and owls. Some dogs kill chickens, and beware skunks and foxes.

If chicks are overcrowded, they may start pecking each other and this can lead to serious trouble as they will peck each other to death. If you see blood on a chicken take a little tar on a stick and completely cover the bloody area with tar. If a chick takes one pick of the sticky, bad tasting tar, it won't peck there again. Give them some lettuce, cabbage, weeds or grass to pick on and be sure they aren't overcrowded.

Your pullets should lay in five or six months. Feed them plenty of whole oats after they are four months old, and change from growing mash to laying mash at five months of age. Don't feed cracked corn as this causes too much fat to form in the hen and makes it difficult for her to lay. This leads to the expulsion of her insides and death from trying to expel the egg through all the fat.

You can also hatch your own chicks. It takes twenty-one days from egg to chicken, and the eggs must be properly incubated, sprinkled with warm water and turned every day.

If you can obtain a setting hen, the job is much simpler. Leghorn hens hardly ever set as they lay almost the year 'round with only about a month to six weeks off to molt (shed their feathers) and have another cluster of yolks form for the next batch of eggs.

Heavy breeds of hens—Rhode Island Reds, Barred Rocks, or Sex-Link Hybrid Cross usually set every year. You don't have to raise a new lot of pullets every year, however, if you have Leghorns. I have had them four years old laying like year olds, with bigger eggs every year.

A wooden bushel box makes a good nest for a setting hen. Fill it about half full of hay or shavings. A big hen can cover up to fourteen eggs. It's best to place the hen on the nest in the dark and cover her with another box. Once a day, let her off for a few minutes to eat and drink and eliminate her waste away from the eggs. The last week, a sprinkle of warm water while the hen is off, might help the chicks to hatch.

Most setting hens will take good care of their chickens, and you won't need a brooder, for they will keep them warm even in pretty cold weather. Don't try to hurry chickens from their shell. Sometimes you may need to expand the hole they've picked, a little, to help them out. But if removed too quickly, they may bleed to death.

You can butcher your cockerels at whatever size you prefer; broiler (2-3 lbs.), fryer (3-4 lbs.) and roasters as large as you want. My husband chops off their head with a sharp axe, in one clean blow. He holds their wings and feet in his left hand and the axe in his right. We prefer doing this to sticking them, as they bleed out well.

Then we dip them into a bucket of scalding water, lift them out and dip them again for a few seconds until the feathers come out easily when you pull on them. Hang the bird up by its feet with strong cord so it is a comfortable height for you to work, and pull out all the feathers. A dull paring knife helps get out pin-feathers, and any hairs can be removed by passing the bird quickly through a flame. I open a cover on my wood range, take a crumpled piece of newspaper, and light it. Then I take the bird by the legs, make a quick pass or two through the flames with him. Then I put several thicknesses of newspaper on the bench beside the sink. With a sharp narrow bladed knife (a fish fillet knife is best), I first cut off the feet by cutting on each side of skin beside the joint just below where the feathers begin on the leg. Then I bend the leg so that it snaps off easily with my hands. Next, I remove the tips of the wings at the first joint. The oil sac, about the size of a half teaspoon measure, is just behind the tail on the bird's back. Carefully cut this out; otherwise, it will give your bird a strong oily flavor.

Then place the bird on its back and cut carefully around the anus. Make a shallow slit in its skin, up to the breast bone. Do not cut too deep—just enough until you can see inside. Then you won't break open the intestines and make a smelly mess.

With your right hand, carefully pull the intestines from the bird. Cut the gizzard free, and open it by cutting around one side. Remove contents. This should be in a sack inside the gizzard, if you use care and don't cut too deep. Have a pan of cold water handy for giblets. Carefully remove liver and cut out the gall bladder (a little green sac on the side of the liver). Be extra careful not to cut into the sac as the bitterness of the gall bladder will ruin the liver. Put the liver in cold water with the gizzard. Pull out the heart and put it in water. Then dig out the lungs and testicles (white, bean-shaped—two inside the back).

Now make a shallow incision in the breast skin and pull out the bird's crop with your fingers. The gullet should come with this along with the windpipe, but you may have to pull it out of the neck separately. Wash all this thoroughly under running water, and then soak them in cold water for a couple of hours.

Place the cleaned chicken in the refrigerator 48 hours before freezing. Do not stack. This allows heat to leave the carcass and makes the bird tender.

Nothing tastes so good as home-raised chicken meat and fresh eggs. Once you raise some, you'll never be satisfied with the store-bought variety again. But chickens are like a garden. They need care, thrive on attention, and suffer from neglect. Don't attempt to raise them if you don't have time or patience to give them proper care.

RAISE ARAUCANAS

I guess it all started when my grandmother decided that baby chicks would make a good present for my first Mother's Day. They would give me more things to mother—as if one active baby boy wasn't enough. She picked Araucanas. As some breeders report, they are more intelligent than other breeds of chickens, tame easily and make good pets. Anyway, ten tiny puff balls arrived by mail from the poultry farm. The chicks were only eight hours old and still shaped like the eggs which had protected them for so long. They were beautifully marked, many with stripes like a chipmunk. One was black with white spots, and the rest were yellow with black stripes on the back or head.

The chicks were quickly settled into a cardboard box in the

kitchen with food, water and a 15 watt bulb. The height of the light bulb from the floor of the box was adjusted until the chicks were neither huddled together under the light nor pressed against the side of the box away from the light. The temperature was kept at about 95 degrees the first week. After the second and third days, clean newspaper was added to the bottom of the box. By the fourth day, the chicks were conditioned to eating the starter mash. Peat moss was then added to the bottom of the box. For grit, I added Hartz Parakeet Budgie Gravel and Grit to their food.

The chicks grew and feathered rapidly and needed larger and larger boxes and more feeder space. When the weather was warm and sunny, we took them outside and placed them in a wire enclosure to let them pick through the grass and chase bugs.

At six weeks they were moved into a new coop my husband built entirely from materials found at the city dump. We kept the chicks in the coop for a week, feeding them starter mash and mixed grain, before turning them out to range for food. They automatically went back to the coop at night.

Now that they have reached maturity, my range birds receive a daily ration or corn, milk, and oats and are allowed a free range through our woodlot and field. They relish all vegetable scraps from the kitchen and are not above stealing any food the dog leaves in his bowl. In the winter, their diet is supplemented with grower mash or laying pellets.

The breeding history of Araucanas is as varied as the colors of their eggs. These tailess fowl with long ear tufts were raised by the Araucana Indians in Chili. These birds were kept in a semi-domesticated state, and the Indians ate them and their brightly colored eggs. When a few of these birds were brought into this country around 1930, breeders wished to improve them in all sorts of ways. Some were crossed with Leghorns to increase egg production; some were crossed with barred Plymouth Rocks and Rhode Island Reds in hopes of producing a dual-purpose breed, and so on, depending on each breeder's preferences. In the process some of the egg coloration was probably lost.

Now, Araucanas are so varied in form that there is no breed standard. Instead, they are divided into three main types: rumpless with ear tufts; rumpless with ear tufts and beards; tailed with ear tufts and beards. Of all these types, the latter is the most common. These chickens can be almost any color or pattern with any type of comb or leg color. Mine all have yellow to white skin, but they may vary, too. Their eyes tend to look somewhat fierce, or hawklike, and may be entirely brown or the normal yellow. Hens weigh four to seven pounds and roosters five to eight.

As a farmstead bird, Araucanas have many points in their favor. They start laying a good number of medium-to-large eggs at about five months and various laboratories have reported that the Araucanas' colored eggs are both higher in protein and lower in cholesterol than white or brown eggs. Each hen's eggs can be identified. One hen always lays the same color egg, which may range from sky blue to mint green, gold to olive, drab to pink. This characteristic gave me some trouble when feeding eggs to my children. My little boy would eat only Big Al's eggs; they were large and pink. If I tried to break a green shelled egg into the skillet, my son started to cry.

Araucanas are very efficient range feeders. Mine relish a lizard or small snake, hunt out beetle larvae and clear out low growing poison ivy by scratching for insects. Don't let them into your garden! They eat plants as well as insects and scratch up everything.

Some hens go broody and others do not depending on what other breeds are in their background. If you want your hens to go broody, do not feed them any laying pellets in the spring. Don't give laying pellets to hens with chicks, either. It can even decrease their mothering instincts to the point where they attack their own chicks!

Araucanas that do go broody generally make good mothers. They are very protective and aggressive enough to handle some predators. When Big Al had ten chicks last spring, all were housed in a five by fifteen foot run made of one-inch chicken wire. One night, a five-foot long black rat snake decided it wanted to eat some baby chickens, and it started to come through the wire. Big Al attacked the snake's head, ripping through the scales. Luckily, the snake was so dazed that it got stuck in the mesh of the chicken wire. The next morning the dog started making such a fuss that we went out to investigate. The dog was after the snake's tail and Big Al was after its head.

The aggressiveness of the Araucanas is also exhibited by the roosters and may be strong enough to cause some problems. We were forced to kill and eat five of our roosters who were hostile towards our two year old child. Fortunately, our sixth rooster does not dance around or jump at people unless they mistreat his hens. Apollo, the mild-mannered rooster, has never had to fight another rooster nor compete for the hens' favors. This may account for his more peaceful behavior, but he still retains enough of his aggressiveness to be a good protector against rats and other small predators.

Whenever a rooster gets a choice morsel of food, he calls his hens and lets them eat it. The mother hens do this with their chicks. They will catch and release many small bugs to show the chicks what to search for. The chicks definitely prefer the bugs to starter mash. They soon develop distinct personalities which they carry throughout life.

Araucanas are generally disease resistant and vigorous fowl, but I do vaccinate for fowl pox at six weeks as it is quite common in my area. I also worm my chickens once each year with Salsbury Wormal tablets. Araucanas do well in our hot, muggy Louisiana summers and do not require any heat in range type shelters on our thirty degree winter nights. They are reported to do well in all climates.

Many hatcheries carry Araucanas. Murray McMurray sells sexed chicks for a very good price. This hatchery is located in Webster City, Iowa 50595. Marti Poultry Farm in Windsor, Missouri ships very handsome straight run chicks. Stromberg's in Pine River, Minnesota 56474, lists two qualities of Araucanas, and they will also ship sexed chicks.

Their colorful eggs, good laying ability, ability to forage for themselves and perky personalities, make Araucanas a fascinating and intelligent choice for the farmstead—or for a Mother's Day present!

LET YOUR CHICKENS RANGE

A gardener who keeps poultry of any kind will derive the most benefit from them by ranging them. Doing so is good for the birds, and if properly managed, good for the land.

Ranging birds is giving them periodic access to pasture. Ducks and geese are almost always allowed range, but chickens seldom are. Out of three serious poultrymen in this area, I am the only one who ranges his birds.

Birds need not range every day; twice a week is a desirable minimum, weather permitting. More than that is fine, as long as the land is not overgrazed. The birds benefit in four ways, and the result is healthier poultry. Healthier poultry means fewer diseases, fewer parasite problems, and fewer losses. And it means better meat, better eggs, and healthier offspring.

First, the sunshine is as good for them as it is for all animals.

It allows them to dry their feathers, and keep warmer in spring and fall. In early spring, especially, hens will lie in the sun for hours, as if to drive the winter's chill from their bones.

Secondly, they get fresh greens, worms and insects. This is important. They sample different plants, chase bugs, and eat dirt as well. In doing so, they are able to correct deficiencies in their diet. Hens are partial to new shoots close to the ground, whereas ducks on the same range will jump to get upper leaves and seeds.

Peculiar to chickens, is their need for dustbaths. When the snow clears in spring, the first thing they do is dig dust-holes for bathing. This fights both mites and lice. By adding wood ashes to their favorite dust-holes, lice can be virtually eliminated. (Mites are held in check by regular coats of creosote or old motor oil on the henhouse floor, lower walls, and roosts.) The control of these parasites is the third way ranging improves the health of your flock.

Finally, the exercise is good for their respiration and circulation. Many of the chronic diseases affecting poultry attack their respiratory systems. The exercise and the sunshine help keep these potential killers from surfacing. Ranging makes for healthier birds in all these ways.

There are possible drawbacks, depending on the point of view. First, egg production per bird may be lowered. Calories expended chasing grasshoppers and taking baths are calories that will not be turned into eggs. Secondly, old hens that have been 'out on the range' are more muscular—which means much tougher as meat. These considerations are offset by the above mentioned advantages.

More birds in good health means more birds fit to lay, and fewer casualties. So, although production per bird may be lower, total production per flock may well be close to the same. There is one final advantage. Since their forage will reduce their grain intake, costs will be lower. This means production of eggs per dollar is likely to be higher for ranged birds, if the range means little or no additional cost outlay.

If you do decide to range your birds, do so with foresight. Ideally, range them in a confined area which can be rotated and gardened every other year. But if not, keep these things in mind: consider where you can afford the space, and make it dry and sunny, rather than damp, shady or marshy. If possible, make it a place which can be tilled. Tilling occasionally will prevent a manure buildup on the surface of the ground—not a healthy situation. If you don't intend to rotate it, make sure that it is large enough so that it isn't a desert by August. Cutting down on the number of days they range is your only control once the fence is up, and chickens, ducks and geese will pick a small area right down to bare dirt in no time.

Fencing is highly desirable. Free-ranging hens and ducks, especially Bantams and Mallards, will nest in the most obscure place they can find. Collecting eggs will then mean looking in thickets, briars, nooks in stone fences, and under all your raised outbuildings. Confining the birds confines the eggs. In addition, birds which are notoriously good mothers, like Muscovy ducks, Bantams, and some geese will set on their eggs through the night, which exposes them to the nocturnal predators—skunk, mink, weasel, owl, and fox. A fenced-in range means the most protected placed for hatching eggs is the henhouse, which can be closed at night, as it should be.

Besides confining egg-laying, an enclosure confines the manure. Bird manure, with its high nitrogen content is much too valuable to be wasted on the driveway, under the woodshed, or on the branches of a favorite shade tree.

Fencing discourages predators (with the exception of hawks, of course). And, a good fence will deter the greatest danger to ranging poultry—dogs. In the past three years, I've lost one hen to a skunk, one drake to a fox, one duck to a mink, and seventeen different fowl to three different dogs. It is a dog's nature to hunt and chase, and a bird in the open is an invitation. If there's a fence, there's hope. Not only is the dog discouraged, but if it manages to jump it or go through it, your legal position is on firmer ground, should you seek compensation. This is no small consideration. A dog can kill a dozen chickens in the fifteen minutes it may take to chase it away.

Finally, fencing keeps the chickens from preying on your vegetables. One duck can easily harvest twenty or thirty asparagus on a morning's walk. Geese know very well how good blueberries are, and any kind of fowl will decimate bean sprouts, young corn and greens of any kind at any time of year. One final note: pigs love chickens, and they can stalk them and catch them as efficiently, if not as gracefully, as any lynx.

If you have snow cover regularly, it is good practice to till the range in fall and sow something so that they will have a good plant cover the following spring. Rye is good, but a mixture is even better. Between it and the self-sown weeds, they'll have a good variety of greens the next spring. This means keeping the birds confined or on another area until the new growth is well-established. One good practice is to turn them out into your garden after harvest. Their usual range will rest, the rye will grow, and the chickens will clean up what you didn't. This is an especially good idea if they've picked their range really clean.

If you have little or no snow cover, the use of the garden as 'off-seasons' range is still a good idea. If you can't, or don't care to, it's still a good idea to keep the birds off the range periodically, to allow it to "digest" the droppings and to recover its plant growth.

The best method is to rotate the range year by year. This fights soil compaction, makes the most of the manure, and prevents intestinal parasites which winter-over in the ground, from re-infecting the flock in spring. Every spring, the birds are given a new area in which to graze, while last season's range is producing food for you or feed for them.

Even if you can't spare a lot of space, or fence, whatever space you can spare will do them a world of good. At any rate, it is certainly preferable to constant confinement in a barn or coop. It will mean that the henhouse needs cleaning less often, saving you work, besides giving you all the rewards that healthy poultry have to offer.

TALKING TURKEY

Now that the holidays are here and our eight broad-breasted bronze turkeys are all waiting for the axe to fall, it's easy to be smug about having raised them to their present enormity. But several years of raising turkeys have underscored once again homesteading's most invaluable lesson: the rewards for one's work are often in direct proportion to the knowledge one has of the subject.

Even though turkeys may sell for as little as $.49 per pound in the supermarket, we find it profitable to raise them. We have no trouble selling a small number for a dollar a pound, dressed, at Thanksgiving. People are willing to pay that price for something healthy and flavorful. We have never tried to market them on a large scale, and undoubtedly many other factors would be involved then, but as a supplement to our other animal-related income, they form an important and profitable link. They need little beyond range, water, and a place to roost, the latter not necessarily being a locked-in house of their own, unless, of course, you're troubled with vermin—a weasel or the like—in which case you'll have to secure them at night. In brief, they're easy to raise.

But turkeys are not just overgrown chickens. The latter know enough to come in out of the rain. Poults (as newly hatched turkeys are called), then, are no brain trust. Can you imagine a bird needing to be taught to eat? Well, that's the way it is with turkeys. Most people throw a few marbles or strips of aluminum foil into the water dish, and sometimes even the feeder, in order to get the critters to start eating. But if you persevere, when you do eat one of your own you'll be in for a surprise as well as a treat—home-raised turkeys taste nothing like those pasteboard commercial birds you find people flocking to buy in late November. Your own birds will taste no less fresh and distinctive than your own tomatoes, fresh eggs, or homemade butter.

Or they should, that is, if instead of raising them in a hospital-like atmosphere, on wire, isolated from all the good things they like to eat, you allow them their freedom. And right here is the key to achieving that raised-at-home taste. Let your turkeys eat off the land. Grass, bugs, apples, whatever lives and can be reached by the turkeys, they'll peck at. (In fact, they'll peck at anything, period. Rubber balls, glass, transistor batteries—remember, these animals are not as bright as, say, your average night crawler.)

In the process they'll eat a wide variety of fresh foods, not only saving you a great deal of money, but producing a flesh of extraordinary flavor. We do make commercial pellets available to our birds at all times, but they don't eat much if they can get outside.

A few cautions, however. They may have to be taught not to peck at a cow's hoof, and they may try to roost on the top of your car. Indeed, in spite of their low I.Q.'s—though come to think of it, perhaps because of it—they can be guided to wherever you want them to go much more easily than chickens. You can herd them. Merely take a long stick or pitchfork and drive them in front of you, holding the stick to one side or another depending on which way you want them to go. Actually, you'll find them following you everywhere. Sometimes they do get in the way; we once nearly sealed one behind a wall we were constructing, much like that unfortunate fellow in Poe's "The Cask of Amontillado."

Of the two major breeds of turkeys, the white and bronze, we prefer the bronze primarily because this variety has more breast meat, though we also enjoy it for aesthetic reasons: a male in full plumage is a handsome bird.

The poults may be purchased from various sources, and should cost about $1.00-$2.25 apiece. We haven't heard of anyone having much success keeping a tom and a hen for breeding, one problem being the size of the tom. In fact, one adult tom raised by a friend for such breeding pretty well destroyed his mate while taking his pleasure. These toms do get big and heavy. Raising them as poults, then, demands about the same procedures as for chickens. Keep them warm—95° at floor level the first week, lowering the temperature (usually by raising the brooder lamp) by 5° a week until they are feathered out or it's warm enough for them. We should caution that they are somewhat more susceptible to cold and drafts than chickens.

If you haven't raised any kind of chick before, keep in mind three principles. One, contain the poults in a circle so that they don't bunch up in a corner and suffocate; two, whatever your thermometer may say, adjust your brooder lamp to the needs of the poults (they should neither be bunched up under the lamp, indicating not enough heat, nor in a corner or edge away from the heat, indicating too much; ideally, they should be pretty evenly scattered about); three, give them enough room (at least two square feet for each poult and at least ten square feet for the mature birds if kept enclosed). One should have feed and water before them at all times, keeping in mind that turkeys are considerably larger than chickens and thus need more feeder space. Our only losses have resulted from human and canine error: we once switched off the barn lights forgetting that the brooder light was on the same circuit; and our Scottish terrier ate one a year ago, mistaking it for who knows what.

Eventually we hope to raise all of our animal feed, but presently we use a Blue Seal medicated turkey starter, switching to a grower pellet after the poults have feathered out. Once they range, however, they actually eat relatively little of this feed, understandably preferring nature's feed to some conglomerate's. We feed all our leftover milk to our hogs, but turkeys should do very well on it also, for our chickens have thrived on it in the past.

By the way, if you have chickens, as we do, or if your property once housed chickens, you may have a problem with what is called "blackhead." The term blackhead (Enterohepatitis) derives from an effect of the disease. The turkey's head loses its reddish color, turning ashen gray—the more severe the disease the more gray the head. Chickens are sometimes affected, but are less susceptible than turkeys and are usually hosts and transmitters for the larger birds. The primary reason for the loss of color in the head is the destruction of vital internal organs resulting in failure of the circulatory system; hence, no blood to the head. More specifically, a protozoa carried by the cecal worm attacks the birds through droppings, affecting poults to 12 weeks of age most severely, resulting in weight loss, droopiness, sulphur-colored droppings, and quite often—if severe enough—death.

The standard control involves the raising of poults on wire cages for the first three months, and then alternating pastures so that the disease will not become established. In our case, we do not fence our birds in, thus we have reluctantly resorted to the use of the chemical Nathiazide, marketed under the name "Hepzide" (available from Northeast Laboratory Services, Maine Poultry Consultants, P.W. Box 262, China Rd., Waterville, Maine 04901). Quite inexpensive, this water-soluble powder should be used whenever symptoms appear, or, as we do, administered as a precaution since we know the disease to be present. Our only loss this year resulted from a hen being kicked by one of our cows. It staggered about for a few days and then decided to give up the ghost.

We usually get our poults toward the end of May and start killing them around Thanksgiving. By Christmas, all are either eaten, sold, or in the freezer. After six months, then, the toms usually dress out to between 25 and 32 pounds, the hens between 17 and 22. The weight of the turkeys makes for some special problems in killing and preparing them, the killing especially being a job for two strong people. Of course, the turkey will have been deprived of feed and water for 24 hours before you slaughter it; just make sure that whatever enclosure he is in does not enable him to peck at hay or feed on the ground.

We have never mastered the method of killing poultry by piercing the brain through the roof of the mouth, so we stick to the good old hatchet. One person holds the turkey with his head on the chopping block (feet in one hand, pinioned wings in the other) and the other chops the head off—hopefully with one merciful blow, although we have had our failures when the bird ended up being bludgeoned to death. Here is where strength is required: once the head is severed, the turkey will do the same twitching as a chicken does; the difference is about 20 pounds worth. It takes a lot of muscle to hold that bird down!

We then immerse the turkey in near-boiling water to loosen its feathers (make sure you hold him down for a count of 60); if you are doing more than one bird at a time, have plenty of water on the stove ready to replenish what the turkeys have splashed about. Then just hang the turkey up by its legs and pull out those feathers. If you've plucked chickens before, but never turkeys, take heart. Their feathers come out neatly and cleanly, leaving a beautiful white skin. As they pluck, so shall they clean; for some reason, turkey insides don't seem as odious as chicken's, the only hassle being the strength required to heave around a 30 pound creature.

A GUIDE TO RAISING DUCKS

Ducks are one of the easier varieties of poultry to keep. They are hardy and their housing requirements are minimal. Ducks are also much more resistant to disease than most poultry. They do have a greater need for water than chickens and this might be a handicap for some backyard growers.

Ducks are kept for both meat production and egg-laying. However, no breed is ideal for both these purposes. The three most common meat varieties are the White Pekin, the Rouen and the Muscovy.

The White Pekin is the duck used in 80% of the commercial meat operations. The Pekin is hardy and does well in confinement. It does not fly and reaches six to seven pounds at seven or eight weeks.

The Rouen is a descendent of the wild mallard and has dark feathers. It is slow to mature, taking six to seven months, but its meat is reputedly very good.

The Muscovy is a distinct race of ducks. If it is crossed with other breeds, the offspring are sterile. They are not well-suited to mass production, but they have advantages for the home-grower. They are the most efficient forager among the domestic breeds, and they are much better brooders. However, the Muscovy's rate of growth is relatively slow. Females will average 5 1/2 to 6 pounds and males 10 to 13 pounds at sixteen to twenty weeks of age.

The best egg-laying breeds of ducks are the Khaki-Campbell and the Indian Runner. In Europe these breeds are used in commercial egg production. They will lay as well as the best strains of chickens. Individual bird records of 300 eggs and more per year are not uncommon. Layer-type ducks need more floor space than hens, requiring about four square feet per duck. They also require about 50% more feed than hens to produce a dozen eggs. However, duck eggs can weigh up to 32 ounces or more per dozen, which is considerably larger than hen eggs.

Among meat birds, White Pekins and Aylesburys are the heavier layers. Flock averages can go up to 120 eggs or higher and they may lay for six months or more. Muscovy ducks usually lay for a shorter period.

For those who wish to raise a small flock of ducks, purchasing day-old ducklings is probably the best way to start. A brooder house suitable for chickens or a warm, dry, draft-free area in a barn may be used for ducklings. Ducks do very well on deep litter, but sufficient materials must be added to keep the surface clean and dry. Straw or shavings are good litter materials, but paper should not be used over the bedding. Young ducklings have difficulty walking on such a smooth surface, and laming is a possibility.

Care should be taken with ducklings while they are in the down stage because at this age they are very susceptible to chilling when wet. Water should be freely available, but offered in a way that will allow them to immerse their beaks without wetting their bodies. This is not always easily done since ducks tend to splash and play in their water dish. Wet bedding will probably have to be removed from the area of the waterer frequently.

After two to four weeks, the ducklings may be transferred outside if the weather is favorable. However, if they are kept in confinement they should be allowed one-half square foot of floor space per bird for the first week, three-fourths square foot for the second, one square foot for the third, increasing up to two-and-a-half square feet by seven weeks.

Ducklings should be offered 22 to 24% duck starter ration in pellets or crumbles on a free-choice basis. A chick starter is satisfactory provided it does not contain any drugs. Ducks are very vulnerable to damage by medicated feeds.

At two weeks of age ducklings should be changed to an 18 to 20% duck grower. Again a chicken feed may be used, but pellets are preferable to mash. Ducks waste a great deal of feed when it is offered in mash form. Your ducks will benefit from pasture. While ducks do not compare with geese as foragers, they still are able to supplement their grain feed with grasses, etc. to a considerable extent. Studies by Cornell University indicate that confinement-reared birds grow somewhat more rapidly than range-reared birds, but require considerably more food per unit of gain in weight. If you are unable to give your ducks pasture area, you can still save on commercial feed by providing them with grass clippings and weeds. They must also have grit and should be provided with an insoluble form from one week on.

If you are going to keep ducks for eggs or breeding, you will not need elaborate housing. A dirt floor covered with litter is fine as a breeder house. Good lighting and ventilation are necessary, but insulation and heating are not usually required—ducks prefer to be outside during the day, even in winter. Nests should be built on the floor with the top and bottom left open. Partitions are constructed twelve to fourteen inches in size and are held apart by being nailed at eleven-inch intervals to six-inch board running along a wall of the pen. A one-by-two inch board is then nailed along the bottom front of the series of nests to make the construction more rigid. Straw or shaving should be placed in the nests, and there should be at least one nest for each three to five ducks. Breeding ducks should have about five square feet of housing space per bird. Plenty of fresh, clean water is essential. There should be swimming water available if your ducks are brooding since it is necessary for them to wet their feathers to provide the proper humidity.

Select your breeding stock when they are six or seven weeks of age. General characteristics to look for are large size, vigor and good carriage. Desirable confirmation is indicated by a strong, clean-cut head, a deep, fairly long body with wide ribs, and a uniform width throughout the bird's keel. You will have to sex your breeders since it is normal to allow a ratio of one drake to each five or six ducks. Ducklings are sexed by inspecting the vent. Sexing should be done under a strong light to make the small penis of the male more easily distinguishable.

Here are some general characteristics which can be used to determine sex in ducks. When mature, the males are usually larger in both body and head parts. The drake's main tail feathers curl forward, and in colored breeds the drakes have the more brilliant plumage. There is also a distinction in the quack of males and females. To the practiced ear, the male quack is softer and less clear while the female's cry is harsher and more distinct.

MORE ABOUT MUSCOVY DUCKS

The Muscovy duck is a native American duck, as it originated in South America. It has been known as the Brazilian, Peruvian, Turkish, Muscovite, Musk, Indian, Guinea and Barbary Duck. For the last fifty years it has been known as the Muscovy Duck.

This duck is different than most other domesticated varieties. It does not quack like other ducks, but rather hisses like a goose. The drake does not have the curled feathers on the tail common to the male ducks of other breeds. The sex is determined by the size of the adult bird. The drake is larger than the hen duck and weighs eleven pounds. The hen duck weighs six to seven pounds. It takes five weeks for the eggs to hatch while other ducks require four weeks.

The Muscovy duck when roasted has a delicious gamey flavor more like wild duck, but without the fishy taste generally associated with wild ducks. For those who like the flavor of a good-tasting, meaty duck without a lot of fat, the Muscovy cannot be excelled.

This duck is very hardy and can be kept outside all winter. They will break ice on water so that they can swim and will spend the coldest night outside sitting on the snow with their head tucked under their wing. They are economical to raise as they forage on grass in summer much like a goose. They have very powerful wings and, unless the outer feathers are clipped, will fly up in trees or roof-tops.

Young ducklings thrive on wet mash. They can be fed duck pellets or the same mash and grain that chickens are fed. However, it is better to moisten the mash, as ducks waste a lot of feed if its dry. They must always have water for drinking.

The Muscovy Duck is of three basic colors—white, black or blue. The white Muscovy is all white with a pinkish or flesh-colored bill, blue eyes and pale orange-colored legs and feet. These dress out for market very nicely with no dark-colored pin feathers left in the skin.

The Black or Colored Muscovy is mostly black with white. The black is predominant on the crest, head, back, wings, and primary feathers. In the colored varieties, the more greenish-black feathers on the bird, the better. The bill should be pink or shell-colored. Its toes and shanks are yellow or dark-lead colored.

The Blue Muscovy has blue and white plumage. It is a cross between the white and the black Muscovy, but it does not always run true to color.

The face in both sexes is bare, bright red, fleshy and carunculated. The male when alarmed or angry erects his crest and turns bright red. This angry countenance combined with his loud hissing gives him a very wicked appearance.

The mother Muscovy is a very good "sitter" and hatches out as many as eighteen ducklings at one time. If the babies are taken away from the mother, she will start to lay again and have a second hatch.

Mother ducks do better if confined until the ducklings have feathered out. If allowed to roam, the mother might take the ducklings out in the damp grass in the early morning where they may get wet and lost as they can not keep up with her. I have lost almost half of one duck's hatch when she left the nest before I expected she would.

A trio of one or two year old Muscovy Ducks will cost around ten dollars and can be obtained from poultry breeders. However, the Muscovy seems to be a very popular farm duck. Perhaps breeders could be bought from farmers in the fall. Sometimes a setting of eggs can be bought in the spring to be set under a hen. To insure the duck eggs hatching, they should be dipped in water the last week of hatching time. This dipping rots the egg shell to help the duckling break out of the egg. When the duck broods the eggs, she wets her feathers before she sets on them for the same purpose.

Chapter 3

Rabbits

Rabbits are hardy animals, and on a small scale, they need very little attention. They require food, water, a cage, light and fresh air. If you should want a rabbit to come to you when you approach it, then you must pay quite a bit of attention to him.

RABBITS ON YOUR FARMSTEAD

We've known many people who thought it would be cute to keep a bunny in their homes as a pet, they successfully litter-trained it, as you would a cat, and then found that "bunny" was chewing up everything in sight, including the legs of a favorite sofa. So, if you are inclined to want to house break these cute friends, keep a careful eye on their activities.

We started out raising rabbits with two doe bunnies from Virginia. They were six weeks old when we brought them to Maine, and they had a hard journey, as they were the back seat passengers in our 1966 VW bug. We traveled well stocked with bottled water, lettuce, and pellets.

When they were six and one-half months old, we bought a buck, Caesar The Great (a very appropriate name). We made a pet of him and he used to come into the house and play with our two Siamese cats. In no time at all, our rabbit herd went from three to 16. We were off to the races.

Within a three-year period, we have gone from three rabbits

to 80, at times. Rabbits are easily raised and bred, if a few-basic rules are followed. Fresh water is vital. Especially in summer, a doe and her litter can drink up to a gallon of water a day. Hard to believe, but true. A dry place to sleep is important too: a wire cage, shavings, hay, or straw are good. Be careful in winter not to use sawdust or shavings for nest boxes, for the new arrivals will not be warm enough. Straw or hay are much more suitable.

The choice of a breed or breeds to raise is important. If you just want a pet, the breed is not so important, but if you are interested in meat production, the larger breeds are the most profitable. There are dozens of breeds from which to choose. Some of the rabbits used for meat are Californians, New Zealands, Chinchilla, or Champagnes. If you are looking for a smaller rabbit for show or fur, the Black and Tan, Lop, and Angora are just a few.

As to the subject of food, rabbit pellets are recommended, due to their high protein alfalfa content. Last winter, grain prices went out-of-sight, and particularly the rabbit feed, so, we started an experiment. We were feeding 18% protein grain to our horses, and started feeding the horse feed to our rabbits. In addition, we gave the rabbits some timothy and clover hay for roughage. This method might not work for some folks, but we have had excellent results.

Before we leave the subject of food: salt must be kept in every cage. Grain stores usually sell rabbit salt rings that are easily hung in the cages. If your doe is having a difficult time breeding, you might also supplement her diet with Calf Manna (25% protein) from Carnation Co., or some Wheat Germ Oil, A-D-E which is great for the rabbit's coat, too.

The approach to housing your rabbit is a personal choice. We started out with wood on three sides, a wire gate and wood floors. This was a very unsatisfactory way to keep rabbits. Even with hay for bedding, which keep rabbits from chewing wood, a few still decided the hay was not enough. When it came time to disinfect the cages with Lysol, the wood would absorb the solution. At that point, we changed to half-inch mesh for the bottoms and chicken wire for the sides, with only the frame in wood. The chicken wire works for mature rabbits; but with the newborns, forget it! We lost many babies due to heads, legs, etc. being caught in the wire. Half-inch mesh wire is the secret. You can also use welded half-inch or inch wire with great success. We built portable cages with two by four boards and half-inch mesh wire, with a wooden top that opens on hinges. Our cages are 30" long × 18" high × 18" wide. Larger cages are nice, but when you have a dozen or more cages, the expense adds up very quickly. All wire cages are the perfect answer to trouble-free rabbit raising.

Depending on the nature of your buck, there is a question of time-limits when breeding a doe. Some breeders leave a buck in a doe's cage for as a little as five minutes; some leave him one week. It depends on the demand of the buck, too. If you have thousands of rabbits that are in need of being bred, then the buck has a limited time with each doe. We have two bucks at the present—both purebreds. One is a large White New Zealand, very excitable and aggressive. We can only leave him with a doe for about five to ten minutes. After that time, the doe gets very irritated with his presence. The other buck is a Black and Tan. His name is Sniffles, and he's a real sweetheart. We can leave him in a doe's cage for two or three weeks and never worry about his manners. One buck has the ability to breed hundreds of does, and the only reason we keep two on our farm is that the New Zealand is a meat producer, while the other produces more showy animals for the Easter market. A reminder! A doe is fertile all but three days out of every month. You might try putting your buck in with your doe for five minutes a day for three or four days, just to make sure she is bred.

Now that you've bred your doe, what happens next? Well, you can expect her to kindle in 28-32 days from the time of conception. She needs a kindling box, which is a very simple item to make. We've made ours from one inch boards to the dimensions of 12″ × 12″ × 6″. This box is filled with clean hay or straw. When the doe is ready, she'll start pulling fur from her belly and forming a nest. This is to allow her babies free access to their source of food. The babies are born without fur, and almost resemble baby mice. If it's the doe's first litter, and she is disturbed at all, she is liable to kill the babies. Be very careful never to touch the newborns for the first week or so. This is enough to make the doe abandon her nest.

Due to the hard winters in Maine, we have found that disease is limited. Three summers ago, however, we had a doe that we had to put away. She showed signs of listlessness and had lost her appetite, though she had kindled a litter the week before. When we examined her, maggots had begun to invade her body. I was shocked at the sight of this poor creature, and immediately took her to our veterinarian. I had no idea of what I had done wrong.

Our vet told me that she had Peritonitis. This was caused by the extreme heat of August, and the afterbirth had attracted black flies. They laid their eggs and we lost the doe. Because of this experience, we make it a practice never to breed in July and August; we have not had any trouble since.

When all those rabbits and bunnies are born, what are you going to do with them? Here are several suggestions you may want to follow. First, the market for bunnies at Easter is fantastic. They are cute and cuddly and sell very well. Second, the meat market for rabbit is very limited in Maine. In Boston, Mass., they are getting $2.79 per pound, dressed. Not a bad market. We eat our own that we raise and they are excellent. (No fat!) Third, we have a beautiful pelt to tan and preserve for rugs, coats, stuffed animals, etc.

There are numerable sources of information on raising rabbits. Your local university extension agent is a good source. Also, the milling companies such as Blue Seal, Purina, Carnation, and Agway are excellent sources of information.

After raising over one hundred chickens to slaughter last year, we have decided to eat rabbit instead. They cost less to raise, and at the end of the line, you have a hide and not a pile of feathers! Good luck to all of you beginners. There are many rewarding hours to be spent raising rabbits.

RABBITS FOR FOOD AND PROFIT

At three months, rabbits are of the frier size for dressing out. They are easily killed by tapping them on the back of the head at the base of the skull, and bled by cutting their throats. If you intend to use the skin, cut around the throat and peel the skin down over the body. Take the front legs and cut off the feet. You will need someone to hold the rabbit by the front legs. Peel the fur over the body, keeping the fur inside so loose hairs don't get on the meat. Peel out the hind legs and cut off the feet and tail. Split up center front from tail to neck and take out entrails; save the heart and liver. Soak in salt water to draw out the blood.

To cut a rabbit in serving-size pieces, cut off the front legs at the shoulder. They are held only by muscles. The hind legs are fastened to the backbone and must be disjointed, and the body is cut into three pieces. Of course, if the rabbit is to be stuffed, it is left whole. After it is cooled it can be wrapped and frozen or cooked.

There are some rabbit growers who have built up a business of selling rabbit meat or dressed rabbit and also rabbit sausage. They will usually take any extra rabbits you have on hand of marketable age. They do not pay much, as they have to make money too, but it is a sure market once it gets started.

If you are interested, you could build up your own business by selling dressed rabbits. Besides the meat, you could sell the skins. The raw skins should be put over a board and hung up to cure.

There is no less expensive meat you can grow then rabbit meat. A three-pound frier can be produced for about 25 cents. It costs about 8 to 14 cents a pound to produce the meat. Rabbit is delicious and tastes somewhat like chicken. In California, where rabbit is king, many people prefer it to chicken. In the super market, frozen rabbit is priced over four dollars for five pounds.

No other meat is as easy, quick or as inexpensive to produce as rabbit and it's easier to dress than chicken. Two good does and a buck will give a family easily 180 pounds of good tasting meat per year.

RAISING RABBITS

How did I get started raising rabbits? I happened to pick up an old issue of *Countryside* magazine, edited by Jerry Balanger, and was reading about others who were raising them for fun, food and profit. I had always been one to picture rabbits as cute, cuddly little creatures who multiplied rapidly and made darling little pets, but that was about all. I was wrong in two areas of my thinking.

Rabbits do not always multiply as fast as people think, and there is much more use for them than just as pets. The more I read, the more interested I became in raising them and having some meat for my table. I had figured on selling them to a commercial rabbit breeder and hopefully to make money by doing that, but soon found out that I had to find and keep my own market. More about that later. First of all I chose an unused building that was just collecting clutter and cleaned it out. Then I bought some rabbit cages from Favorite Manufacturing Co. of Pennsylvania, and bought some wire and made some of my own hutches. The hutches I prefer are 30″ × 18″ for the breeding does. I have six hutches that are 24″ × 24″, used as holding pens for the fryers and the ones I am going to sell or keep for breeders. The hutches have pans under them so I can keep all the manure for the garden. As rabbit manure is one of the best fertilizers you can put on your ground, I didn't want to waste any of that because I garden organically and need all of the natural fertilizers I can get my hands on.

After getting my rabbitry building ready, the next thing was to get some rabbits. I had already written to several rabbit breeders and made arrangements with a man in Mechanic Falls to buy two

New Zealand White Does and one New Zealand white buck. I would recommend that any breeder who has any ideas of raising rabbits start with good stock. Your offspring will only be as good as their sires. I was fortunate to get good rabbits and was anxious to get started. Let me mention at this point that I had read a lot of articles, and talked with different people who were also raising rabbits. I didn't totally start in the dark, but what I didn't know would have filled more books than what I did know. And while the books and articles are helpful, I doubt that the rabbits have read them because each rabbit is different! I was anxious to get my rabbits bred and was waiting for the arrival of lots of little bunnies so I could begin selling them right away!

I was in for a surprise. I had thought that rabbits were always in heat. That is not true. There may be some rabbit breeders who would disagree with that statement, but I know from experience that it isn't so. Rabbits have a heat cycle like most other animals, and while it is true you can get them to breed quite easily in the spring and summer try it in the fall and winter! Rabbits have a 16-day heat cycle and if you are lucky, you can get them bred. Sometimes a rabbit will just huddle in the corner of the buck's hutch and growl at him each time he approaches her. Let me mention here too, be sure to always put the doe in the buck's hutch and as soon as he has serviced her, take her out. Do not leave them together! This is a good way to have a good buck get castrated, for a doe will tear him to pieces in just a matter of minutes if left together. I know there are people who do leave them together, but they are taking an awful chance. If the doe is in heat, she will squat down in the hutch and put her tail up over her back, the buck will mount the doe immediately and fall off to his side as soon as the job is completed. It will all happen in a matter of minutes.

After the doe has been bred, put her back in her hutch and leave the rest up to her. In an average of 28-32 days the doe will kindle. Be sure that she has a nest box put in a couple of days before she is due to kindle. I usually put shavings and some hay in the nest box so the doe can arrange them to suit herself. She will usually pull some fur to go along with what you have put in. When the bunnies are born they do not have any hair and can freeze to death in just a matter of minutes if the doe has them on the wire or hasn't pulled any fur. Usually if the doe has them on the wire, or pulls no fur, there is a reason for her doing so. Maybe something disturbed her about the time she is getting ready to kindle, or she has no milk. If a rat or a cat, or dog is present in the rabbitry, she may have them on the wire. If that happens wait about three days then breed her back. If she has no milk, unless you have another doe that is about to kindle on the same day and has proven from experience that she has lots of milk and is able to care for extras, cull the doe that has no milk, as she will never be any good as far as a producer goes. With the price of feed the way it is, you cannot afford to keep a rabbit that doesn't prove up. As soon as the doe has kindled and has gotten out of the nest box to eat, check the litter to be sure that they are okay. If there are any dead ones remove them and then put the nest box back where it was and leave the doe alone because she will resent having her family touched, even by her owner. I always make sure that a nursing doe has plenty to eat, and I also give her warm water to help her. Make sure that all rabbits have a salt lick in their hutches at all times. Rabbits need salt just like other animals do and if deprived of this, it can be a contributing factor in causing them to eat their young. I feed my rabbits Coarse 16 which is a dairy ration. Why? Because Coarse 16 is high in protein, and quite a bit cheaper than rabbit pellets. Also I have found that there isn't so much waste, because there isn't so much dust in the feed as there is in pellets. If there is too much dust in the feed, rabbits can develop the snuffles, which are very hard to get rid of once they get started. My rabbits have done very well on Coarse 16 and it doesn't seem to take as much to get them up to butchering size. The little bunnies should stay with their mother until they are about two months old and then it is time to cull them out, by picking the ones you want to butcher, and the ones you want to keep for replacement breeders, and the ones you want to sell for breeding stock and for pets.

There is quite a bit of work involved if you want to succeed at keeping rabbits. Rabbits need to be kept clean. In the wintertime it is hard to keep the rabbitry clean, but it is just as important then as it is in the summer. Urine and manure will freeze to the hutches and collect if the pans are not cleaned. I clean my pans out everyday. I have to thaw them by using boiling water and then washing them clean. I find that using newspaper to line the pans also helps. It will decompose and go right back into the soil with the manure, so there is no need to worry about that.

I also give my rabbits plenty of warm water to drink when it gets cold. Don't worry about the cold, as rabbits have plenty of warm fur and can stand it a lot colder than they can the heat and humidity of summer. I have seen it down to − 10 in the rabbitry itself, and it never seems to bother them.

When the time comes to butcher, I hold off feeding them for 24 hours so there is less feed in the stomach or bowels. I do give them all they want to drink, and this serves mentioning here also. Always look for a rabbit that will drink plenty of water, for this is a sign that the meat will be much better. Rabbit meat has a lot of moisture in it, and is much better if the rabbit will drink hardily from the time they start drinking to the time they are butchered.

There is no easy way to butcher. That is never pleasant, but if you are raising rabbits for table meat that is the only way you are going to get them onto the table. I admit that I had a hard time killing my first rabbit, but if you know how, it will help. There are many books on the market that will give you some help, but like me, you will have to work out the system that works best for you. I hit the rabbit behind the ears with a blunt object—a good thick stick works best. Make sure you hit the rabbit with a hard, swift blow. Then remove the head immediately and let the rabbit bleed out. I remove the head with a pair of pruning snips and then hang the rabbit on a gambrel hook to let it bleed. I then proceed to remove the front feet, at the joint. You can find this by bending the front foot and cutting it off right at that point. Don't cut any higher—there is good meat on the front leg and you don't want to waste it. I then cut off the tail right at the rump. Then take a sharp skinning knife and slit the skin down the inside of both hind legs until you come to the vent area. Be careful here. You don't want to rupture the bladder or the bowel tract, and you will if you jab your knife too hard and don't know what you are cutting.

After you have the two hind legs skinned, you can pull the fur off the rabbit just like a glove. After you do one or two rabbits you will find a method that works best for you and you will find that you can do it quite fast. Rabbits are much easier to do than chickens. After the skin is off, you have to take the insides out. This sounds awful, but after you get the hang of it, it comes quite easily. I cut around the anal area very carefully and then pull the insides down toward the chest cavity. Be careful that you don't puncture the stomach or any of the insides. I usually cut the bladder off and remove that separately so that it won't get punctured. After you have the insides out, cut off the liver, and take out the heart. Some people will find these very tasty, and even if you don't like them, they make an excellent treat for a cat or dog after they are cooked. Now you can cut up the rabbit. I do this by cutting off the hind legs just at the place where they join onto the body of the rabbit. I also cut off the tail at this point. Put the pieces of meat into the pan of cold water so that any fur that has gotten onto the meat will soak off. Do not let the meat soak longer than 15 minutes, as rabbit meat has a tendency to soak up water. Cut the back joint off right at the two front legs. There will be a piece right around the neck area that is waste; throw this away with the insides. You should have seven pieces of rabbit meat in all. After this has cooled and been thoroughly washed you will notice that the meat is white. Unlike chicken, there is no dark meat on the rabbit. Any recipe that can be used for chicken will work for rabbit. I cook rabbit meat about 40 minutes at 400° if I am baking it. Be careful you don't overcook it, or it will dry out on you.

I think that with a little practice and a lot of patience, you will find that rabbit meat is delicious, easy to raise, easy to butcher, and most of all, easy to eat! Rabbit meat is low in fat and calories, which makes it good for people who are on special diets and trying to lose weight. You will make mistakes the same as we all do in raising them. There will be times of discouragement. But when you eat that first rabbit you have raised yourself, you will find like I did, that it was well worth every minute of work and effort. And when you see how well your garden grows from the manure, and see the wonderful produce you get, you will be like the rest of us homesteaders who are raising rabbits—you will be hooked for life!

BARNLOFT RABBITS

Colony breeding in rabbits has come under close scrutiny in recent years as a method of increasing the production and decreasing the cost of the farmstead rabbit. However, the experiences of those involved seem, curiously, to have been largely negative. It is my belief that this attitude is due to the improper construction and management of the rabbit colony.

Since the origin of the conventional rabbit breeding system, rabbit raising has always been accompanied by the relatively high expense of individual housing and cage furniture. Added to this is the investment of time involved in feeding and caring for singly housed animals. After all, it usually takes as long to feed and water a rabbit as a pig, and the value of grown pig averages ten times the value of an adult rabbit.

The simplicity of colony breeding on the farmstead has drawn increasing interest. There is no individual breeding, feeding, or expense involved. The principle of colony breeding decreases the importance of the individual animal, but individual attention to your animals should have an even greater emphasis in the colony.

When I became interested in group breeding experiments, I was put off by the descriptions of disease ridden, inbred, unhealthy animals. Most information available on the colony system will tell you that rabbits, when raised as a group, will interbreed uncontrollably, spread disease throughout the colony, dig out the enclosure, and generally take advantage of the opportunity to become furry delinquents.

And guess what? All this is perfectly true! But what nobody seems to realize is that the critics of colony breeding apparently advocate the absence of any human intervention once the colony has been originated. This is patently absurd.

The first step in successful colony breeding is to get your rabbits up off the ground. This is where most group breeding systems begin to break down, because a doe rabbit is among nature's most capable excavators. Added to this are the factors of weather, predation and contamination from diseased wild animals. Getting your rabbits off the ground will solve half the problems experienced by most colony breeders, and it can be accomplished in a variety of ways.

The barnloft is the ideal place on most farmsteads. With a minimum of work the unused area in your loft can be converted into usable livestock space. However, there is endless opportunity for the use of other buildings that meet the minimum requirements for the profitable colony.

In all cases, floors should be wooden, two inches thick if possible, with tightly fitting boards at least eight inches clear of the ground. If a roofless enclosure is used, walls should be solid one to two inch thick wood, at least four feet high to take the place of building walls (which your rabbits will nest against). The top should be enclosed with strong fencing, and at least a partial wooden rain-breaker.

If using the loft, you will want to make a couple of additions to your loft floors to keep waste material from trickling down. After partitioning off your colony area, make the following adjustments:

Cover the floor with a sheet of heavy plastic. Cover this with a two inch layer of sawdust. On that lay some smooth surfaced material, such as masonite, but clean, burr-free tin roofing, drilled and slanted formica, and many other materials are acceptable. Your rabbits will dig down to this surface, so make sure it is smooth and conduct periodic checks to make sure the rabbits aren't digging or chewing through it. On top of all this, place at least four inches of sawdust.

In the raised colony, no furniture is necessary other than food and water containers. A large salt block should be hung at least six inches off the bedding, from a non-metal support. When figuring the amount of floor space needed a safe general estimate allows at least one square yard for each adult breeding doe in the colony, and half that for the herd buck.

To begin operations, fill the enclosure with whatever bedding material you prefer. We used straw, but soft wood sawdust, hay, ground cobs and dry chopped fodder would all work. The rule of thumb is: the more bedding, the better. Pile it as high as you can, keeping in mind the rabbits Houdini-like fame for escape.

This bedding will serve as home, nursery and social center for your rabbits. The does will burrow in to have their young in a warm corner cavern. The bedding will be trodden, stomped, chewed, urinated on and generally degraded. Keep plenty in there, and clean it out and change it at least three times each year (not much when you consider the weekly cleaning of individual cages).

Place two sexually mature does (not bred, since they need some time to settle down and call the place home before breeding) in the enclosure, feed them, and other than periodic health checks, leave them alone for two weeks. After this, pick a healthy, proven buck, and drop him in with them.

This is where most sources conclude with ". . . and that's it—harvest your fryers six months later . . ." But if you want a successful colony, you're going to have no choice but to devote some time to it.

Although the work time factor involved with the colony is still far less than that for individual housing, the colony requires daily attention to grow and prosper. Aside from the necessities of proper food and water, available fresh and daily, your animals must receive constant medical attention. The one great fault of the colony is the rapidity with which disease may spread. But this can, in almost every case, be avoided through careful attention.

Look for all the signs of disease daily in every animal you can catch without tearing up the bedding. The ones hiding will be available tomorrow or the next day. Handle them frequently and they'll be less skittish. Look for wounds, runny noses, diarrhea, discharges, sore feet, debris or pimples in the ears, hair loss, and general thriftlessness. For treatment, refer to a good book on rabbit diseases and administer medicines promptly. If the condition is contagious, it is imperative that the rabbit be removed immediately and quarantined.

With attention, the inbreeding problem is no problem. Keep records, tattoo your animals, and simply remove any that you don't want to breed. It is generally unwise to allow more than one buck to a colony. The exception to this may be two like brothers from the same well documented litters. If your stock is standardized, your fryers will be too. But it is impossible to keep records with two unrelated bucks breeding indiscriminately. Fryers are butchered at eight weeks old, so you'll have no excuse to leave young ones in unless you want them to reproduce.

A word about varmints: Buy mouse and rat traps and keep them set outside the colony enclosure. Pour a little ammonia down open rodent holes and fill them in with dirt. This seems to move them pretty quick. Keep varmints trapped out and keep feed covered. Make sure that there is a closeable door to keep larger predators away from your colony.

I'm not going to go through the usual spiel: ". . . two does produce eight young which are breeding age in 170 days which produces . . .", because these are "ideal condition" predictions, and your results will vary tremendously. Let me simply say that with planning and attention, a colony with five producing does can let a family of three have rabbit for dinner three times a week, year in and year out.

And you'll only have to face that nemesis of the homesteader— the cage cleaning blues—three or four times a year.

Chapter 4

Sheep and Goats

If you have ever watched small sheep or goats frolic, and wondered what it would be like to raise them, the following sections should prove useful in helping you make your decision.

HO, SHEEP, SHEEP!

Sheep offer a great deal to the Maine farmstead. A small flock can produce meat for the family table plus extra income from wool and surplus lambs. Spring lambs can be ready for the freezer by fall and grown on hay and pasture with a minimum of grain. Although good fences are a necessity, housing needs and labor requirements are small. Sheep are easy to handle and are one of the lowest priced farm animals available. Sheep have been a popular and profitable Maine farm enterprise in the past, and chances are good that they will return to favor again.

We first became interested in sheep three years ago. After building our barn and filling it with hay, we were ready to start a small flock. Although sheep can often be picked up cheaply at livestock auctions, we knew we didn't know what to look for in an animal and feared relying on pot luck. A casual remark by friends led us to Paul Patten's small farm where we bought a four-year-old Dorset ewe and two yearling North Country Cheviots for $55. He shared many of his experiences with us and we trusted in his good advice. We picked up our "girls" in early November, after

Paul was certain that they had been pasture bred, and put them in the loafing area of our barn with our three goats. The goats quickly established the rules of the house and everyone got along fine.

We clipped away the wool around the ewes' udders and tails and watched for their udders to enlarge, indicating the approach of lambing. "Madam Dorset" was the first and presented us with a ten-pound ewe lamb in late December, followed by the two Cheviots with single and twin rams in March. The ewes were put into a separate pen about a week before lambing where we kept a close watch on them, for this is when they need their greatest care. Fortunately all eleven of our lambs have come normally, without help from us.

As soon as we found the lambs, we wiped them dry with old towels, then shortened the umbilical cord and dipped the end in iodine. Old burlap bags were tacked around the pen walls to reduce drafts. A little help was sometimes needed to show the youngsters the way to their "milk bar." A little warm milk in the palm of the hand and on a finger started the lambs sucking and instinct quickly took over. Within a few days we clipped the tails back to about an inch stub with a sharp knife. Blood spurted momentarily but a small rubber band worked well when double wrapped around the skin that had been pulled over the tail end. A little iodine helped and the stubs healed over quickly. We fed the ewes about a pound a day of a 14% protein coarse-fitting ration for a month prior to lambing and until they went out on pasture.

Our hay ran low that first spring. So soon as the new grass was two inches high, we turned our little flock out on pasture. Our permanent fenced pasture was still in the planning stage—and the fun was about to begin. We soon realized that sheep are impatient wanderers, feeding together in a loose group and moving almost constantly. In a morning's time they had explored every part of the ten-acre field that surrounds our house, barns and gardens. They were not interested in the woods, the young balsam-fir Christmas trees, or the fruit trees. Periodically they returned to the barn for water and to lie in the shade. It didn't take long for them to discover the strawberry bed and the young peas, and it became nip and tuck to see if the plants could grow faster than the sheep could mow them down. As the lawn greened up they kept the young grass and wild white clover mowed down. Sheep are far easier led than driven. They quickly learned to respond to a pail rattling a few handfuls of grain. A call, "Ho, sheep-sheep!" and sight or sound of the pail would bring them on the run. They didn't mind the rain and I think it helped keep their fleece bright and clean.

We fenced our permanent pasture as soon as possible. We set cedar posts at ten-foot intervals and stretched a four-foot high woven wire fence around a two-acre area at a cost of approximately $200. Although it was more expensive, we preferred the vertical stays at six-inch intervals to keep the small lambs inside and to reduce chances of them working the fence with their constant reaching for grass through the wire. If dogs become a problem, we can stretch one or more strands of barbed wire above the present woven wire. A 55-gallon drum cut in half and set up in a wood cradle provided a serviceable water trough.

We spent a day with our friend Paul, learning to shear with an electric clipper, and borrowed them to continue our education at home with our three ewes. Our efforts produced about 22 lbs. of wool which we kept until we found buyers.

That fall we built a sheep shed in the pasture. It was an 18 ft. by 24 ft. pole shed with a partition and raised floor for hay storage, a feeding rack, and a loafing area with a six-foot entrance. The spruce, pine, and aspen lumber came from logs cut in our farm woodlot; shingles, paper, nails and custom sawing cost about $250. This bar holds about two tons of loose hay—a winter's supply for about eight ewes.

We purchased a half interest in a nice looking purebred Dorset ram and put him in with our four ewes for November and December. Unfortunately, we were late in butchering our three ram lambs in January, for by then they had lost much of their early fall fat. The carcasses ranged from 25 to 40 lbs.

Our new lambs arrived during the first two weeks in May, five months after Sandy's services. He had apparently bred them all during their first heat after his arrival. We had not finished shingling the sheep shed the previous fall, so we brought the sheep back to the main bar and penned them in a large box stall for lambing. Our youngest ewe, Lomb, presented us with a single ewe, and the three older ewes all twinned for a crop of three ewes and four rams. We finished shingling the sheep shed roof and put the flock back into the pasture in early May as soon as it provided grazing.

We sold two ram lambs at four-to-five weeks of age for $25 each. Sale of young lambs offers an opportunity to split up mixed twins, leaving the ewe's milk to boost the growth of a promising ewe lamb. At this age the lamb's future qualities are beginning to show. (Raising a young lamb to market weight at six months of age will add only another $10-15 of value.) We sheared the older ewes in early June and later sold the wool to hand-spinners.

Pleased with the lambs that Sandy had sired, we decided to use him again with the older ewes as well as with his three spring ewes. He came in early December so we are expecting our lambs this May. We are interested to see how well these younger ewes manage, bred at eight months of age. We like our Dorset and Dorset-Cheviot crosses. The fleeces are good, the ewes are good mothers, the Dorset horns make a convenient handle, and the lambs develop well.

Adding up our income and expenses for the past two and a half years, we show the following Table 4-1.

Table 4-1. Economic Profile for Raising Sheep for One Year.

Income			
Wool	22 lbs. @ .65 -	$14.30	
	23 lbs. @ 1.50 -	34.50	
	4 lamb fleeces -	16.00	
			$64.80
Lambs	5 market @ .80/lb. -	138.00	
	2 live -	50.00	
			188.00
			252.80
Expenses			
Feed - 600 lbs. grain			45.00
Investment			
Flock - 3 ewes		$55.00	
Sunbeam Shearmaster		68.00	
Fencing - 4 rolls		200.00	
Sheep Shed		250.00	
			$573.00

Our income has paid the feed bill, recovered our initial investment in the flock and shears, and bought half the fencing. We have not included the value of our hay, pasture, and labor since they do not represent a true outlay of cash. The three lambs we have used ourselves are included in income at market values. With a good lamb crop this spring, we may be able to·recover the remainder of our investment in fencing and the shed this year, leaving us with our equipment and facilities fully paid for as well as a flock of seven ewes.

Looking ahead, we plan to fence in an additional four acres. This will allow us to rotate and improve our pastures and possibly increase the flock to ten to twelve ewes. We hope to maintain the quality of our fleece, selling to hand-spinners—a good market for a small flock with limited wool production. We prefer selling our surplus meat lambs at weaning age, taking advance orders from individual customers. We are not interested in building our flock beyond the limits of our hay, pasture, and optimum markets. Regular soil tests help us keep track of soil fertility which is built up by periodic additions of ground limestone and clamshells, wood ashes, and manure. Hay and pasture improvements pay dividends in terms of healthy and thrifty livestock.

Whenever possible we feed garden residues to our sheep. Corn stalks, bean vines, squashes, cucumbers, cabbage leaves, etc. quickly disappear when thrown over the pasture fence. In late fall after the harvest and when pasture is sparse we let the sheep range over the farm, cleaning up left-overs wherever they can find them. Mangels, pumpkins, cull apples, and apple cider pomace make palatable high-moisture supplemental feeds through the fall and winter.

So far we have had no disease or worm problems. We use Hoegger's Worm Compound, a mixture of botanicals available from Hoegger Supply Co. in Milford, Pa., mixed with their grain to keep our goats and sheep free from these troublesome pests.

The suggested reading references may provide a background to sheep raising, but we have learned more by watching and working with our small flock. Shearing and slaughtering take practice but can readily be done at home with a small investment in the proper tools. A store-bought lamb is no match for the taste of chops, legs, and roasts from the animals you raise. And given a disease-free flock, lamb is one of the lowest cost meats you can produce . . . and you can do everything yourself.

LAMBING TIME

The sheepman should plan to spend extra time with his flock during the lambing season. For the shepherd this is his harvest time. You should do everything possible to save your harvest. Saving the extra lamb will mean more income for the sheepman.

Breeding dates should have been recorded, so you will know when to expect lambing. Three weeks before lambing time the sheepman should start getting ready. The wool should be clipped from the ewes udder, flank and up to the dock. Why? For sanitary reasons and to keep the lamb from eating wool, a habit which often leads to death because of "wool ball" formation in the lamb's stomach. The ewes should be made to exercise. One way to accomplish this is to feed some of the hay outside away from the barn. The ewes will travel to get the hay, and as a result get the exercise they need. A medicine chest should be set up with the following items: iodine (for use on navels of young lambs), soap, sterilized cotton cord (keep in sterilized bottle), lubricant, disinfectant, paper towels, rubber gloves, hand sheep shears, methiolate, dosing syringe, nursing bottles and nipples.

Sheepman should have all equipment needed for lambing ready beforehand. The area that is being used for lambing should be clean, disinfected and dry. Individual lambing pens, at least five by six feet, are a must. Portable pens are convenient and easy to make and at least one pen should be available per five ewes in the flock. The ewe and new born lamb should be kept in this pen for 2 days.

Feeding Ewes and Lambs

The last six weeks before lambing the ewes should be getting at least one pound of grain per day, plus hay and all the water they will drink. Grains that can be used are 14% dairy feed, or 16% dairy feed if hay is poor. You can mix your own grain mixture made up of:

60 pounds oats		50 pounds oats
30 pounds corn	or	30 pounds corn
10 pounds soybean oil meal		20 pounds soybean oil meal

For the first two days after lambing the amount of grain mixture fed to the ewe should be reduced by half. After this the ewe should get one pound of grain if hay is good and one and a half pounds if hay is poor. This feeding should continue until the flock goes out to pasture.

Milk is the best feed for lambs, but this should be supplemented with grain as fed to ewes and the very best hay. The lambs should be fed in a creep feeder by themselves. This creep feeder should be built in a sunny corner of the barn, by spacing slats at a width which allows the lambs to enter but excludes the ewes.

Signs That The Ewe Is Going To Lamb

It is very difficult for a person to tell just when the ewe will lamb. If you have kept records on breeding dates this will help in estimating the lambing date. A good shepherd watches ewes. The udder will fill, the ewe may keep to herself, and her external sex organs will swell and have more color. When the ewe shows labor she should be allowed to take her time. She will strain and get up and down. When the front foot of the lamb appears, then the other, then the nose everything is normal.

If this position is not showing or the ewe has been in prolonged labor, she is going to need assistance. Before you assist the ewe, disinfect your hands and cut your fingernails. Sanitary precautions are very important when helping a ewe. The first step is to determine the lamb's position by gently entering the ewe's vagina with your hand. Before entering the ewe your hands and arms should be washed with soap, then rubbed liberally with oil. If you have any cuts on your hands, disinfected and oiled rubber gloves should be used. If the lamb is in the wrong position, you can assist the ewe by gently moving the lamb into proper position.

Among the wrong positions are: head and one front leg forward other front leg back; head forward with both front legs back; head back with both front legs forward; lamb lying on back; or backwards. In all these positions except backwards, the lamb should be pushed back with the hand and straightened to normal position. In case of the lamb coming out backwards the lamb should not be turned, but delivered in this backward position. When doing any of these jobs the shepherd should be very careful, working slowly and easily since the uterus of the ewe is very thin and can be damaged very easily.

Care of New Born Lamb

Following birth of the lamb iodine should be put on the lamb's navel cord immediately to prevent infection. Lambs often have a film over their nose and nostrils which should be removed at once to prevent smothering to death.

A lamb that is lifeless at birth can sometimes be revived by giving artificial respiration, slapping the side of the lamb, pumping the front legs, blowing into the mouth, and picking the lamb up by the front legs and spanking sharply. The shepherd should help this type of lamb get some mother's milk immediately. This can be done by setting the ewe on her rump and laying the lamb on its side, put a teat into the lamb's mouth, and squeeze milk into it until the lamb gets a taste and begins to nurse on its own. It is extremely important that the lamb be nursing and getting milk from the mother and that the mother is owning the lamb. This is why the ewe and lamb must be isolated in the lambing pen for a few days.

What should you do about a disowned lamb and how can you try to get the ewe to own the lamb? Proper nutrition for the ewe will help. Try to get the ewe to recognize her lamb by smell by smearing some of her milk on her nose and then on the lamb. Tie the ewe in a small pen so the lamb can nurse frequently. It might also help to put a dog in the next pen to make the ewe more possessive of her lamb.

If you must feed an orphan lamb from a bottle, keep the bottle clean and warm the milk to body temperature. Give a small amount of milk every two hours, and in three to four days increase the amount of milk and the time between feedings. A good mixture to use is 13 ounces each of evaporated milk and water with two tablespoons of corn syrup.

Shortly after lambing a ewe's udder should be checked for mastitis and she should be observed to see if she is over producing milk. If a ewe is over producing she should be milked out by hand or she should be held so other lambs can nurse her. It is also important to make sure that the lamb is taking milk from both sides of the udder.

Lambs born during cold weather may become chilled. One of the best ways to revive a lamb that is badly chilled is to rub it briskly with a cloth and wrap the lamb's head and body by rolling it in a dunlap bag. With your hand, hold the lamb's head and after removing your hand there will be an opening so the lamb can breathe. Always try to get some milk into the lamb before wrapping it. When the lamb has revived, it will come out of the bag.

All of the extra time you spend with your ewes and lambs during this lambing period is well worth it. The good management practices you follow could mean a greater return on your sheep enterprise.

GETTING YOUR GOAT

"I just can't handle it anymore," my neighbor was saying, his face weary and tense. "I've given my goats the best but they walk all over me. They get through the fence at least twice a day and they've girdled my best apple tree. I've reached the limit—tomorrow they go to the auction." Here was a man who loved goats, as many people do, but he had been driven to the breaking point. I understand completely, for I'd been through the same thing. When you're trying to have a working relationship with an animal, and it won't cooperate, it can be a very frustrating state of affairs.

I've run into quite a few people who keep goats and say, "Oh, I never keep Daisy tied up or penned in. She's like a pet, just runs around and does her own thing." This can work out very well if you have no vegetable garden, no flower beds or ornamentals, no fruit trees, nor any close-by neighbors who do. For added safety you should live away from a road. (Even if you swear Daisy only crosses to munch on that thick patch of wild raspberries when no cars are in sight, accidents do happen and drivers aren't usually prepared for crossing livestock.)

But most of us who want to keep productive farm animals also like to raise vegetables and fruit trees, and it's asking too much of a roaming goat to resist the temptation of your luscious-looking cabbage patch. I tried giving two of my goats a short period of "free play" every evening before milking time, while I was going in and out anyway and could keep an eye on them. They were really good for a couple of weeks, sticking to an area around the barn where no destruction was possible. Then one day I turned my back for just a minute too long, and they headed right for the front of the house—established as forbidden territory—and pruned back my peony and geraniums severely. The temptation had been too great, and I really could blame only myself.

Having raised goats for a few years, I must offer this advice to anyone considering the endeavor: decide first how you will confine the animals when they're out to pasture, and make sure you can afford it. Properly confined, your goats will be reasonably happy and so will you. If they never have the opportunity to plague you with such antics as knocking over the lumber pile and pulling clothes off the line, they won't miss it. And you won't be constantly battling, making yourself frustrated and miserable.

Different goats have different temperaments, just as people do. Kids, of course, are generally more rowdy, which is part of the reason they're so lovable. They don't cause much damage when they're little, and it seems like such harmless fun to let them use the farm equipment to practice their jumping. But as they get bigger it's not so cute, and can be downright obnoxious when things get overturned and broken! Happily, goats do settle down considerably as they get older and are bred; they certainly aren't likely to jump as high once their udders are weighed down with milk. But why let the kids get bad habits you'll only have to break later?

When you're shopping for goats, temperament is an important point to consider. It's helpful to know the parentage of the goats in question for many reasons, and temperament is one. Such qualities as noisiness and aggressive tendencies do run in families. Do the kids you're selecting come from a line of extremely vocal goats? Goats can range from being very quiet to screaming and bellowing constantly; it's nice to find a happy medium. Some will be stubborn, obnoxious and more persistent in trying to break fences, while others are more mellow and seem eager to please. This can make a great difference in how you get along. You can still do well buying an animal whose background you know nothing about, but you must be sure you can recognize physical characteristics and be confident that you can cope with any personality problems should they arise.

Once you've established that you want to keep your goats properly confined, the next question arises—what method to use? If you don't want your goats breaking out twice a day as my friend's did, you must offer something better than a slipshod, rickety structure that merely serves as a challenge.

Ideally, a four to five foot high fence of two inch by four inch woven wire is the answer. If you're planning to keep several goats and you know you'll be sticking with it, this kind of fence will pay off. It is costly, no doubt about it; many find it prohibitively so. But if you decide to take the plunge, you'll have made an important improvement on your farm.

If you have a choice between the type that is spot-welded where the wires meet, or the type with little bits of wire wrapped around every juncture, the latter will prove stronger. Avoid chain-link style fence mesh, as this is easy to crawl under or climb over. Whatever type you decide on, be sure to attach the fencing on the goat side of the posts. If you fasten it on the outside, the fasteners may become undone as the goats worry the fence rubbing and scratching themselves. Run the mesh around the posts at the corners for added strength.

If you have access to free slab wood at a sawmill, don't mind the work involved and aren't too concerned with appearance, a picket-type fence could be built that would keep anything in.

Electric fencing seems to work for some goats and not for others. A strong charge is very important, as we found out; when we tried using a weaker charge our goats just didn't respond to the shock produced. Several strands of wire are needed. You won't be able to string the lowest wire too low, as it would short out on the grass, so kids will be able to crawl underneath. This type of fence works best for full-grown goats.

Right here is a good place to comment on barbed wire. I would not recommend this at all, as it is really a hazard. It's too easy for an animal to get badly torn up, and torn teats especially are no fun to deal with.

Some people will not consider tethering, while others never use any other method of confinement. If accompanied by some measure of intelligence, I think tethering can work out well. There are always stories of goats getting tangled on their ropes and strangling to death, and I know of many close calls. It does happen! It's just plain stupid to tie any animal where there is anything in reach to tangle on; I would only consider tethering in a clear, grassy area. Use a swivel on the end of the stake, and another on the goat's end of the rope or chain to prevent twisting. Goats can exhibit amazing strength, as we found out many times, so make sure that stake is well-driven into the ground. Fifteen feet of leeway is a good amount to allow; constantly moving the stake will not be necessary.

Make sure, if you're tethering several goats, that they're far enough apart so as not to get tangled together. If you have a large, clear area this method can work out well for your animals. It would be wise to invest in enough fencing for an exercise pen where they can get together, run around and play.

Tethering involves, in the long run, more attention on your part than fencing does. You'll have to move the stakes every day and there will be times when the sun is too strong to leave your animals exposed in an open area. A good fence, that your goats won't bother trying to get out of, will mean less overall work once the initial job of erecting it is done. Your animals will be happier, too.

Why are some goats a pleasure to milk while others put up such a fuss that you begin to dread doing chores? I think it's important to establish right away a good attitude towards milking. A milking stand makes the whole affair a lot more pleasant. My husband built one out of scrap lumber, following the standard plans that seem to be printed in every goat book. A couple of weeks before my does were ready to kid for their first time, I began feeding them while they stood on the stand. They loved to hop up and soon became accustomed to slipping their heads in the stanchion to gobble their grain. When they kidded and milking time arrived, getting them to cooperate was no problem, and there never was a kicked-over bucket of milk. It's important that you genuinely enjoy milking and convey that attitude to your goats.

Your goats must learn, too, that hoof-trimming, grooming and hair-clipping, worming and other routine management practices are all part of their lives. You should establish a grooming routine, and be firm about completing any operation you begin. No goat I've ever met has wanted to stand still while having its hooves trimmed. You might tie yours on a short rope in the barn for this, and firmly take hold of the foot. You can be kind and gentle, and still show who's boss. Don't let an uncooperative goat get away with half a trimming job or without its worming medicine. If you let your goats walk all over you, they'll get in the habit, and every hoof-trimming session will be a dreadful experience.

These ways of dealing with goats are useful with any farm animal, and can go a long way towards making your relationship a satisfying one. A happy animal is a pleasure to see, but that happiness does not have to be at the expense of yours. Your goats can be content and cooperative, and if you can maintain that, you won't be likely to find yourself reluctantly shipping them to the auction and returning to store-bought milk.

THE FARMSTEAD BUCK GOAT

There's only one thing that books on goat husbandry are sure to agree on; keeping a buck goat is a nuisance. If you have just a few does, they say, you certainly don't need to keep a buck. He's big, he's smelly, he's hard to handle and he needs a lot of exercise. In fact, they often go so far as to advise prospective goat owners to first sniff out the nearest good buck and then buy a doe of the same breed—just to avoid having the old fellow around.

But obviously, some people keep bucks, and some of them, like me, even like having them around.

We had not intended to get a buck until an older woman friend of ours became ill and was unable to take care of her older animal. He had been quite a pet for her, and she was anxious to find a home where he would get the attention and care he was used to. She sold him to us for much less than he was worth, and we've been pleasantly surprised by how well he has worked out for us. We certainly think that a well-cared for buck goat is not a nuisance but an asset to any homestead.

As I see (and experience it), there are at least five advantages to owning your own homestead buck.

He Insures That Your Does Get Bred

For us, becoming as self-sufficient as possible is what homesteading is all about, and to pay someone else to do what we could do for ourselves is simply not necessary.

As anyone who has a female goat knows, it's often not just a simple matter of paying someone else to have the goat bred. It is up to the owner himself to know when his doe is in standing heat (that short and often elusive period of time when the doe will stand for the buck's services) and to get her to the chosen buck in time— hoping that the buck's owner is at home and that the buck is available for service (what if he's already had his quota of does for the week?).

Moreover, many goat breeders think that the male of the species can bring a reluctant female into heat; at any rate, all would agree that he is the one who knows best when the doe's time has come.

So, if you own your own buck you have a heat detection specialist in residence. You know that he is available and that there will be no hassles in getting the doe to him, either the first time or later if she should need to be re-bred. And, you don't have to pay someone else a stud fee! His breeding your does insures, barring pregnancy and kidding problems, that there will be milk (and butter and cheese and cream) and meat (should you desire it) on your table for the coming year.

He Can Pay His Own Way and Then Some

If you own a purebred buck, you can have a public stud service. The usual charges for breeding a doe run from $10-$50, depending on the bloodlines behind the buck, and his ability to sire does who produce a lot of milk and show well. And, if you have extra space in your barn and someone can be home to check on the animals, you can board the harder-to-breed females for about a week at $.50-$1.50 per day.

We have a purebred French Alpine buck and live out in the sticks. We don't even have a phone and have advertised just by word of mouth, but our buck has always serviced at least 10 does a year at $12.50 per service. This goes a long way in paying the feed bills.

It is important to check each incoming doe to make sure she is in good health—you don't want your boy getting sick! Also, do not use him too frequently, no matter how much he seems to be enjoying it. All those sniggering jokes about the "prodigious sexual appetite of the goat" are true! Three times a week is plenty for the mature buck, and be sure to give him a sprinkling of wheat germ over his food during the breeding season—it helps keep his strength and vitality up.

He Can Work Around Your Place

Goats are herd animals and the male goat's natural role, in addition to insuring the continuation of the species, is protector of the herd. This instinct has not been bred out of the domestic goat and it is good to know that the herd sire is alert and ready to defend. While somewhat limited in his abilities (his lack of horns prevents his being able to gore predators), the desire and power are still there.

One of our does, who has the run of 33 acres, was attacked by a pack of stray dogs and our buck managed to drive them off before they did irreparable harm to her.

I also think that the does like having a buck around, just as hens like having a rooster; he has a calming influence over them. Perhaps they know he protects them.

He Can Be a Good Companion and a Source of Fun Around the Homestead

The buck goat, once acquired, is bound to be one of the more permanent members of the homestead. He will certainly not be eaten and can thus be allowed to become something of a pet if his owners are so inclined. I think that giving the buck plenty of attention and affection makes him easier to handle because it increases his desire to please. From my experience, I would have to say that bucks are naturally more affectionate than their female counterparts; they are not quite so capricious and seem to have an innate desire to please, or at least receive attention.

Our buck loves to be on hand for wood-chopping—the "thwack" of the axe splitting wood seems to excite him tremendously; he dances around on his hind legs, tail wagging furiously. Maybe it's reminiscent of his ancestral call to battle. He and I often go for walks in the woods together (he probably thinks of it as "dining out"). I've found that whenever I'm in his company, I can approach deer much more closely than I can when alone. Sometimes I sit in the sun with him while he chews his cud. I find him a good influence;

he has both a zest for life and a calm acceptance of it, which I find a helpful reminder.

These advantages are all well and good—but don't bucks smell? Aren't they hard to handle?

When I see a buck that smells particularly bad (this is usually accompanied by a dirty, waxy coat) or is ill-mannered instead of friendly and alert, I know that he probably has an owner who doesn't give him much attention. Usually these smelly, pushy fellows are housed in small dark sheds and have little contact with human beings or other goats.

Yes, a buck does smell—but only during breeding season. This is due to his obnoxious habit of spraying his urine over his beard and belly. But if the buck is bred only to the does in his herd, his breeding season and resultant odor, go into a decline after he has serviced his last doe. A buck can be kept relatively odorless by bathing him with a pine-scented soap, if weather permits, or if not, spraying him with a dilute bleach solution. Also, it is now possible to have a buck permanently "deodorized" by removing the musk gland on top of his head. Since it is the odor from this gland that causes the scent of the urine to linger in such an unpleasant manner, this surgery can make a real difference and is done routinely by many breeders.

In any case, make sure the buck has fresh air and adequate room to exercise. When our buck had to be housed on dry bedding for a month due to an injured foot, he simply reeked, because he had nothing to do all day but eat and spray himself. I am convinced that bucks will do a certain amount of spraying in any case, but when they become bored or are denied social contact, the amount of spraying increases markedly.

Incidentally, we house our buck with the does. We milk in a separate area, and have never had a problem with the buck's odor lingering in our milk.

A buck that is well-fed and has the room to exercise and contact with human beings should not be hard to handle. Our buck comes when he is called and leads easily. A buck should not be permitted to develop bad habits, such as butting people. We've found that telling our buck "no" firmly but patiently is usually enough to restrain him from doing something which is annoying to us. If he persists in misbehaving or is particularly rambunctious, we give him a quick squirt with a water-filled squirt gun. This is infinitely better than either hitting him or pushing him away, since these just make a buck disagreeable. I would recommend buying a buck kid and working with him frequently; if you do buy an older animal, be sure that he doesn't have any bad habits or that you have a lot of time and patience.

Chapter 5

Les Eggleton

Pigs

Can money be made by raising and breeding pigs? Which do you butcher, and which do you keep? If you raise them for the meat, how will you go about butchering pigs? The answers you seek are just a few pages away.

BREED THAT PIG

It's the time of a year for that little piglet you bought to be a not-so-little pig. I am sure your experience has grown as the pig has grown, and perhaps you are now torn between filling your freezer and keeping the animal for breeding. I have learned the hard way that there are a number of factors which must be considered before attempting such a venturesome challenge as breeding.

How to tell a good-looking pig from a worthless scrub is as difficult for the beginner as it is for me to describe. Visiting a commercial hog farm or an animal auction where a number of pigs can be seen and compared would be of help. The following description from *Management and Housing for Confinement Swine Production* describes potentially good breeders thusly: "Select gilts that seem to show desired frame and length (a long lean frame for bacon production); sound in their feet and legs; walk wide front and rear; have at least 12 good teats and possess adequate muscle (as shown by a minimum of backfat and a natural thickness over their top and loin)."

It is essential that your gilt be not overly fat as this will hinder her exercise during gestation, make labor more exhausting, and cause her to be clumsy around her little ones after farrowing. A pig force-fed fat producing foods and destined to be "off its feet" by fall (as used to be the practice in New England) is far too fat. Most commercial hog farmers choose and separate their breeding stock by five months of age and feed these animals less carbohydrates and starch.

Looks ain't everything. If possible, find out a bit about your pig's parents: the number of pigs per litter, the breed and body traits of both parents, and some indication of their overall thriftiness (weight gain/quantity of feed). These inheritable traits not only indicate the quality of your pig but will help determine the choice of boar you use.

What this boils down to is that you should be concerned primarily with traits which can be genetically passed on to the sow's offspring. As you can see, the size of the litter your pig came from is a less important factor than the thriftiness and structural traits of the parents. Commercial farms pick only the best looking and fastest growing animals of each litter for breeding purposes. The importance of sow quality in a breeding enterprise becomes a matter of economics in the pig market, where good looking piglets command higher prices and good shoats raised for slaughter demonstrate higher meat quality.

The disposition of your gilt can make a lot of difference during and after farrowing. A mean sow can be a nightmare to work around. Picture continuous harrassment by a 450 lb. sow, such as lunging at my arms and face as I am outstretched to examine the suckling pigs.

There are two major types of swine: the shorter and fatter "lard" type, and the longer, leaner "bacon" type. Poland China, Duroc Jersey, Hampshire, Chester White, Berkshire and Spotted Poland China demonstrate the lard characteristics. Yorkshire, Tamworth and the new breeds mentioned in Raising Hogs In Maine by Brugman and Goater—Minnesota No. 1 and No. 2. Montana No. 1 or Hamprace—are the bacon type. Pig breeds are continually being improved as commercial pig farmers search for the perfect hybrid. The trend now is away from the lard type to the more commercially marketable bacon types. Unless you are raising pigs, specifically for breeding stock, there is no need for keeping to purebreds. Crossbred pigs many times will demonstrate a hybrid vigor characterized by faster, growing, more thrifty pigs in larger litters. What this means is that if the boar sire of your pig was a Yorkshire, for example, more hybrid vigor will be likely to occur when you breed your pig to a Hampshire boar. Such three-way cross demonstrates more hybrid vigor than a two-way cross or a cross between two pigs of the same breed.

It is a good idea to use a good looking, purebred boar on a grade or a cross bred. This will insure genetic improvement from litter to litter. If possible, get the breeding record of the boar you plan to use. Not only will this give you some idea of his potency, size of potential litters, and thriftiness of offspring, but it can reveal any genetic defects present. If no record exists, talk with other farmers who have used the boar. I bred my two sows to a good looking Yorkshire boar last fall and found later that there were at least four ridglings in the two litters (ridglings are male pigs on whom only one testicle descends into the scrotum, making castration impractical).

In this Maine climate, your sow will need more than a three-sided building as an abode. A well ventilated, yet windproof and warm structure is needed. My barn would make a good windtunnel for NASA. It is a 19th century windblown antique that has seen better days. Faced with the prospect of wintering over my two sows and having no money to do it with I chose the following plan: I built two 10′ × 10′ shed roof structures, utilizing rough boards and spruce logs over four inches in diameter for studs. For the roof (which had the regular barn roof over it) I stacked bales of hay, leaving one section open about two feet by five feet for ventilation and a clean-out door. I put a window and a door in each structure for me and a door to the outside pasture for the mamas. To insulate these walls I nailed rough boards on the inside to a height of four feet and stuffed this space with hay. This was all done very quickly and cheaply. The floor was hay-covered dirt. Heat generated by bacterial action in this hay added extra warmth to the enclosures. There were some important extras which made tending the pigs much easier. Pig proof locks were put on each door. (I'm sure that by now you have some idea what I mean by this). Four strong spruce rails were secured eight inches to three and a half feet off the ground running the pen's length and dividing the space into two parts. This allowed the piglets to "creep" feed without competition from the sows. This homemade farrowing pen worked rather well. I was able to tend the little ones and visit with the mamas without risking retaliation. The pig proof lock to the pasture enabled me to keep the mamas out and examine the piglets.

As the bedding became shredded I would shovel it out the ventilation hatch and replace it with fresh hay. Last year I farrowed a litter in February with no trouble at all.

When you keep a pig all year round it is necessary to have a bit more room for pasture. Gestating gilts and sows need plenty of exercise. The pens I used were roughly 3000 square feet per sow. This enabled them to root in relatively clean soil. It is a good idea to rotate pig pasture to prevent massive bacteria buildups in the soil. Pigs can get from twenty to thirty percent of their feed requirements from good pasture, but the pasture should be well drained. A water cover creates an aerobic condition which encourages a group of bacteria that produces ammonia, carbon dioxide, hydrogen sulfide, and methane gas. These gases are what give pig farming a bad name. Bacteria in dry pasture oxidize organic wastes into carbon dioxide, water, and plant nutrients.

A note about fencing here. No doubt your pigs have escaped at least once. For a large area I recommend woven wire. The posts should be set three feet deep and placed eight feet apart. Electric fence works well if your pig is trained young and if the fence is kept free from grass and weeds. Run one strand of electric wire about nose high (the pig's nose that is) on posts fifteen feet apart. This might mean moving it once or twice a summer for growing shoats. Somehow my pigs have pulled the disappearing trick with every fence I've tried. It's best to check them every once in a while.

If I were to list on paper all the foods that pigs could eat, and then soak the paper in water, they would eat that too. Nevertheless there are certain basic foods which breeding pigs need for health and vitality. Sows will ingest fetal pigs to maintain their own bodies if not given enough proteins, carbohydrates, roughage, minerals and vitamins. Not only will good quality feed increase the sow's chances of having a large, healthy litter, but it will also enable the mama to produce a continuous supply of milk after farrowing. The availability and cost of these five basic foods will determine the best and most economical way to feed your breeding stock.

Protein feeds, such as fish, dairy products and tankage (packing house by-products) contain the highest percent protein per pound. Vegetable proteins such as linseed meal, soy bean meal, cottonseed meal, and dried ground legumes or legume pastures (clover, ladino, alfalfa, rape) can greatly reduce the amount of costly animal proteins needed. I have read in various sources that gestating gilts need from 12 to 15 percent protein. Common sense tells me that 15 percent would be a better figure, and commercially mixed grains in this part of Maine contain 15 percent protein. Garbage from restaurants according to Brugman and Goater in *Feeding Hogs* contains about 17% protein. I have fed my sows cooked alewives (an ocean fish which breeds in fresh water streams) during the spring when they can be gotten by the bushel.

Grains provide carbohydrates used for metabolism and stored as fat. Four sources of carbohydrates which can be grown in Maine are corn, oats, barley and potatoes. Keep in mind that during the winter months pigs need more of this basic food to maintain their body temperatures. Clover, Ladino, rape, alfalfa pasture will provide vitamins, minerals, proteins, and carbohydrates and will therefore cut down on the amount of these feeds bought commercially.

Brugman and Goater suggest the use of stabilized iodized salt, bone meal (or dicalcium phosphate), oyster shell flour (or ground limestone) in equal amounts as a mineral feed. Nancy Bubel in her article "The Pig Report" in *Mother Earth News No. 17* mentions a mixture of ten pounds steamed bone meal, ten pounds limestone or wood ash, and five pounds salt. Commercial feed mixtures are designed to be well-balanced in this respect. I would recommend using a salt lick in addition to any mineral mixture.

Vitamins are essential to life processes. It is sometimes necessary during the winter months, when the young pigs are unable to get outside to root to supplement their diet with a spoonful of cod liver oil (rich in vitamins A and D). Symptoms of a vitamin deficiency would be a lack of vitality and a pale skin color.

Feed mixtures should vary depending on the quality of pasture you have and the specific feeds you plan to use.

The book *Practical Animal Husbandry* mentions a mixture of six parts corn or ground barley, four parts oats, and two parts tankage. The author recommends two and a half pounds of grain for each 100 pounds of live weight. This later formula seems a bit easier to comprehend. A few conversion factors mentioned in this book might be of use to you. Ten pounds skim milk replaces one pound of tankage. Weight of corn on the cob feed should be increased by one quarter to allow for weight of cob. Forty-four pounds of garbage a day are needed for a pig 150 pounds or over.

This spring I purchased 5000 pounds of seed potatoes for $30 and used them with commercially mixed grains. A diet of good clover pasture, potatoes and mineral mix can be used for brood sows until roughly a month before farrowing, at which time they should be given a regular gestating sow ration. While I was busy every weekend tending the fire of scrap wood under a 55 gallon drum

Table 5-1. Feed Ration Profile.

Ratios with Pasture		Dry Lots
If barley and tankage are available		
Barley	80	65
Oats	16	10
50% tankage	2	3
44% soybean meal	2	7
alfalfa (or clover)		15
	100 lbs.	100 lbs.
If corn and tankage are available		
Corn	75	58
Oats	20	15
Tankage	2	4
Soybean meal	3	8
Alfalfa		15
	100 lbs.	100 lbs.
If barley and fishmeal are available		
Barley	80	66
Oats	16	10
60% Fishmeal	2	3
Soybean oil meal	2	6
Alfalfa		15
	100 lbs.	100 lbs.
If corn and fishmeal are available		
Corn	75	59
Oats	20	15
Fishmeal	2	4
Soybean oil meal	3	7
Alfalfa		15
	100 lbs.	100 lbs.

From a circular printed by the *Cooperative Extension Service* of the *University of Maine.*

of cooking potatoes, Dr. Holmes of Belfast was using a steam cleaning device to cook his potatoes in a fraction of the time it took me. By the way, although there are fewer vitamins in cooked food, pigs can better utilize it in this form. Another interesting phenomenon is that if you provide "self feeders" containing all the necessary food ingredients, the pigs will balance their own diets. What an easy way to do it!

Now back to animal husbandry. Gilts should be at least six to eight months old and weigh 200 pounds before being bred. A sow or gilt will come into heat about every 21 days and remain in heat from two to four days. Most experts recommend breeding on the second day of heat, and then at least once again during the remaining heat period. The vulva will swell and redden during this period. Sometimes penning a boar near a female will bring on the estrus period. If possible, bring the boar to the female's quarters (picture us driving a 400-pound boar in our Volkswagon bus over to see our mamas). Don't keep the boar with the female after her heat has ended. After fertilization (there is no easy way of telling), take good care of the sow for three months, three weeks and three days. A few days before the due date the sow will start to build her nest. Our's would break off branches of brush from the pasture and haul it into the farrowing pen. Your initial reaction is to help them by stuffing the pen with hay. This may look bucolic but it does more harm than good. The hay creates a pocket around the sow, which during farrowing becomes a deathtrap for the newborn. Twenty-four to 48 hours before farrowing, the vulva becomes swollen and milk will start to flow from the udders. The sow will become uneasy and groan a lot. At this point, check the animal frequently. If you are fortunate enough to be there when the little ones are born, pick the burst embryonic sack off and cut the umbilical cord about an inch from the abdomen, place them beside the mama's teats, and wait for another to be born. Labor lasts anywhere from two hours onward depending on the size of the litter. We found that sometimes pigs would be born five minutes apart and sometimes we had to wait 40 before another arrived.

Piglets are born with about three days worth of iron in their bodies. If there is not sufficient iron-containing dirt present (as when the ground is frozen) there are four things you can do. Place dirt in the pen, add iron supplement to their feed, rub an iron solution on the mama's teats, or inject an iron solution into the ham muscles of the piglets.

At about three days old, the piglets will start exploring the area in the farrowing pen. Place a container of moist, high-protein food in the creep feeding section of the pen, away from the sow. I recommend Baby Pig Starter, a special mixture of grains with milk and antibiotic additives or use regular Sow Chow and add the milk additive yourself (it's cheaper). The sooner these little ones start eating the creep feed the better.

Although I have never done this, experts recommend clipping the tusks of piglets within the first week to prevent damage to the sow's teats. Between two and six weeks of age the male pigs should be castrated. Castration is necessary for two reasons: pigs can breed at early ages, and the meat of boar pigs killed after seven months of age has a rather strong odor and taste. I have found the easiest time to castrate piglets is at about three weeks of age. There is a government pamphlet which describes the how-to of pig castration that will be of great help the first time you do it. With a little reading and perhaps some brandy fortification you should do fine. Leave the castrated barrows with the mamas for a week or so before weaning.

The piglets can be weaned anywhere from three to eight weeks. This depends on what kind of facilities you have and the health of the animals. During the winter months litters farrowed in unheated buildings should probably suckle six to eight weeks. On the other hand, if the sow has mastitis or is having trouble coping with the constant harassment by the piglets you might want to wean them earlier. Lock the sow into the farrowing pen and restrict the amount of grain and water she gets. Put the little ones in a pen out of hearing distance from the sow. In three days or so the mama's milk will dry up and the piglets will have adjusted to the change.

Mama knows best when it comes to piglets, though it is inevitable that piglets will get cut or stepped on. It is possible to raise pigs on cow's milk after three days of age; however, I wouldn't recommend this. We tried raising one that had a huge cut on its back. We bottle fed it every two hours day and night and watched it die slowly. Since then we have left wounded pigs with the mama and have had good results.

Within a week after weaning, the sow should come into heat again. You can breed her again or wait another 21 days. Young sows (six months to a year old) should not farrow more than one litter per year. Mature sows can handle two litters six months apart. Some farmers breed their sows three times a year. Plan the weaning time to coincide with good weather conditions as well as the pig market. It is no fun caring for piglets in January that you are unable to sell. Usually people buy pigs in the spring or early summer.

A final word: the questions I had concerning hog management seemed to blossom at the worst possible times in the most impossible situations. Do your homework before starting such a venture. If complications arise, call the local vet for advice. Despite any mistakes I made in the beginning, the experience of breeding pigs was well worth the time, money, and emotional commitment.

TO MARKET, TO MARKET, TO BUY A FAT PIG . . .

If buying a pig were as simple as the old nursery rhyme made it seem, it would be a lot easier for the small holder who's out to put chops and bacon on the table. But before the auctioneer says, "Get on the seats, boys," the prospective pig-finisher had better have his act together.

Around here, they tell the story about the farmer who needed a pig to use up extra milk from the family cow, and arrived at the community sale just as the last pig of the day was being sold. Rather than return home empty-handed, he bid the pig in at top dollar, and for the next six months he fed the pig, which didn't gain an ounce. Finally admitting defeat, he took the pig back to the sale and stood agape when the pig entered the ring and the auctioneer remarked, "Well, it looks like ol' Little Bill is back again." The lessons the Little Bill's of the world teach are costly ones to learn, but the small farmer out to buy a pig or two doesn't have to be at the mercy of pig jockeys and sale barn slickers.

Feeder pigs are available from many sources, including community auctions, graded and association sales, feeder pig shows, and local farmers. At graded sales, pigs are sorted by weight and quality, and farm-fresh pigs are worth a premium because they will be less stressed and less likely to be diseased from exposure to other hogs. The farmer seeking one or two pigs is at some disadvantage, as producers generally try to raise pigs in large enough numbers to attract commercial finishers, but there are pigs to be had and at bargain prices if the farmer is willing to spend some time shopping for them.

Pigs in small bunches sell down the line because the big operators don't want the trouble of blending bunches, such as boar pigs, some high-backed pigs, and small (20 to 35 pound) pigs during cold weather. Our local vet put himself through college buying ruptured pigs, pigs with downed ears, and junk pigs; but such problems are expensive to correct even when done by a veterinarian and are best not attempted by the farmer. Pigs with a cough, breathing difficulties, swelling in the joints, nasal discharge, bent or twisted noses, a dull, listless manner and eye, or exceptionally rough haircoat should be avoided. A series of terms used to describe faults in pigs being offered for sale might include:

Walnut: a knot or swelling on the animal, an abscess; some can be corrected by lancing, but others may indicate internal abscessing.

Mama's pig: a runt pig that has been raised as a bit of a pet, it is said that mama's pig only comes to town for one of two reasons, mama needs a new pair of shoes, or she can't get the rascal to grow.

Slop pig: the same as mama's pig.

Slaughter ticket: the pig in the ring must be sent to slaughter because of disease or injury.

As is: anyone who ever bought a used car knows what this means.

Rough: may mean just a case of lice or mange or a slight touch of pneumonia, but be damned careful.

Thumping: unnatural or labored breathing.

Stale: a pig with some age on him, and small for his age.

Tail-ender: a poor-performing pig often sold off a group, small for his age, but not always stunted—tail-enders and stale pigs will often perform well when placed on their own, given a change of rations and a bit of medication.

Hump-backed: a pig with bowed-up back; can indicate worms or may be structural; if they have good chest capacity and show some frame they will grow into meat animals, but may be a bit slower.

Heavy on one side: an enlargement of one side of the scrotum, indicating a scrotal rupture.

Ridgling: a male with an undescended testicle.

Good pigs are always worth the money and to paraphrase the commercial, they take some of the worry out of being a hog-raiser. At graded sales, pigs will be inspected by a government agent who will sort them as to weight and quality before they are offered for sale. At community sales or when buying at private treaty, there are visual indications that can guide the buyer. Pigs should exhibit bright eyes, an alert manner, shiny haircoat, and free and easy movement. They should have deep, wide chests, a large skeletal frame with heavy bone, good length, and adequate muscling, although the current trend is to a flatter muscled animal that will grow to heavier weights and still grade high. Purebred or crossbred pigs should perform equally well for the home finisher and it is important that they show good weight for age, 40 to 60 pounds at eight to 10 weeks of age.

Terms used to describe thrifty pigs which are doing well include:

Fresh as spring water: a catch-all phrase that will vary from region to region.

Daylight: a term indicating length of leg and frame; when the pig stands up, you can see plenty of daylight underneath.

Stretch: another term for length.

Big headed: just that, an indication of frame.

Deep sided: indicating a large body cavity, conducive to good feed consumption and growth.

We have produced feeder pigs for nearly 20 years, and if there is any consistent pricing guide, we have yet to find it. But two old rules of thumb say a 40-pound pig should be equal in worth to 100 pounds of market weight hog, or a per pound price should be in the range of 1.5 to 1.8 times the per pound price of market hogs. It is best to set a price ceiling before setting out on a buying expedition, lest you succumb to the sale barn sillies. With the selection made and the pigs bought, the new pork producer's work is actually only just beginning.

Before bringing the pigs home, they should be vaccinated for erysipelas. Some states require that feeder pigs be inspected by a veterinarian before being transported.

Upon arrival home, the pigs should be placed on a bulky ration (approximately 20 percent of the grain should be some form of oats and of a moderate protein level, say 14 percent). Too high a protein level, or a too-hot ration, could trigger scours. The pigs will be stressed by being moved, and if they have come out of a sale barn, they must be considered exposed to several diseases. Isolate them from other hogs on the farm, and give them medicated drinking water for three to seven days. There are numerous sulfa and electrolyte products, as well as Terryamicin, available for medicating drinking water, and I prefer the latter because of better palatability. To improve the palatability of sulfa products, flavored Jello can be mixed into the water. Water medications are to be preferred to feed medications, as the first few days after transport the pigs may be slightly off feed.

Pigs should be considered wormy and carrying lice at the time of purchase, but should be given a week or two to settle down and make adjustments, before being treated. Korlan and Tiguvon are lice-control products that are easy to use, can be used on pigs of all ages, and can be used year-round without stressing the pig. Recommended wormer varieties include Piperazine (can be used in feed or water), Tramisol (available in a pellet or for mixing in water), and Atgard (to be mixed in ground feed). The latter two are considered most effective with a slight nod to Atgard.

Castration is a fairly simple procedure to learn, but should be tried only after observing a vet or experienced neighbor and doing it the first few times under their watchful eye. Never attempt to castrate a boar with a scrotal rupture, or whose testicles appear to be heavy on one side. There are two ways to castrate: 1) make an incision over each testicle low on the scrotum (to promote drainage), squeeze on each side of the testicle, and then draw it out and sever the cord or 2) hold the pig by the hind legs, head down, and then make the incisions in the flesh over the now-bulging testicles between the hind legs, and then follow the same procedure as above. The wounds should then be sprayed with iodine and the pigs observed closely until any swelling goes down.

Housing and equipment for the small pig producer need not be fancy, and fortunately most of it can be homemade. A feeding and watering trough can be made at a very nominal cost, by salvaging an old hot water heater tank from the local dump, cutting it in half, and welding simple cross members across the ends (to give stability) and across the top (to keep the pigs from lying in them). Housing need not be elaborate, but as there is a touch of bulldozer in every pig, it does need to be stout.

A simple A-frame or Smidley type house will provide shelter for from one to four pigs and can be made with salvaged lumber, plywood or corrugated tin. There should be some kind of insulation in the ceiling to cut down dampness from respiration, and it should have solid floors to keep down drafts and assure a clean dry area for sleeping. A four by six foot house will easily accommodate one or two pigs to market weight, and a six by eight foot house should handle up to four head. A pen of similar size should be placed in front of the house and should have a slotted or solid wood or concrete floor to eliminate problems with mud and reduce feed wastage.

A solid-floored pen should be sloped so urine will run off and manure can be scraped out under the gates and scooped up for the compost pile. A slatted unit, with two by six-inch flooring with one-inch spacings is recommended, and should be elevated on concrete blocks so manure can be raked from beneath it. Red worms can be introduced beneath such a floor, to help break down wastes. Such units should be erected with doors opening to the south and some means of shade for the hottest part of the day. Pigs will perform best when grown out in the Spring and Fall and housing should be tight enough to provide protection from the damp and drafts, and keep them up out of the mud. A wet, dirty pig will perform poorly and runs a much greater disease risk.

With the pig well-housed and off to a good start, it now becomes a job of monitoring his health and keeping him moving (growing). The best health advice I can give is to just go out and look at him. Listen to him. Observe for coughing, thumping, sneezing, listlessness, gauntness, and diarrhea. Diarrhea (scours) can often be treated by administering something as simple as buttermilk to change the pH and bacteria content of the gut, and rather than feeding antibiotics in the ration, many believe it is best to prevent disease problems through careful stock selection; clean, dry housing, a plan of good nutrition and parasite control, and use of products such as lactobacillic acid in the rations, with the use of antibiotics only to treat specific problems.

The small holder should be guided in his feeding program by pig type—the framier, flatter-muscled types, for instance, can be pushed hard to desired slaughter weight. You can choose between giving him commercial, pelleted complete rations; a grind and mix program, or self-feeding grain, (and hand-feeding him a protein supplement). Mineral supplements, when needed, are generally offered free choice. Starting at 40 pounds, a pig will generally consume eight to 10.5 bushels of corn and 90 to 125 pounds of protein supplement, depending on the slaughter weight fed to.

With only a few pigs to feed, the pelleted complete rations are often the most economical choice, as they require no special processing or handling or storage. Pelleting gives an estimated five to 10 percent increase in feed efficiency. Commercial protein supplements are available with a protein content of 35 to 45 percent, and can be blended with ground grain (grinding supposedly enables the pig to make the fullest possible use of the grain). These blended rations are generally sold by the ton, with the buyer paying the cost of grinding, mixing, and delivery, as well as the cost of the components. Another choice is to buy whole grain corn to self-feed, and then hand-feeding one half to three quarters of a pound of protein supplement per head, per day.

Milo can replace corn in the ration, but it must be ground, and vitamin A supplementation may be necessary. Wheat can replace up to 50 percent of the corn or milo in a ration, but it must be ground coarsely, for if ground too fine, it presents a palatability problem. A pig can use surplus milk and some table scraps, but these are difficult components with which to build a balanced ration and should be considered as a plus to the regular ration.

After the start-up period, pigs in the 40-to-70 pound range should be fed a ration with a 15 or 16 percent protein content. An example of a 16 percent ration would be corn blended with a 40 percent protein supplement at the ratio of 3.43 pounds of corn to one pound of supplement. The vitamin, mineral, and protein content of commercial supplements is inspected and enforced by law, and there is very little difference in the 40 percent (or whatever) supplements offered by different companies. The commercial supplements are complete; whereas SBOM (soybean oil meal) must have vitamin and mineral premixes added to formulate complete rations with grain. Seventy-to-150 pound pigs should be fed a 15 percent ration, and 150-to-240 pound pigs should receive a 14 to 15 percent ration.

The above rations should produce a feed efficiency of three to four pounds of feed to a pound of grain. Feeding protein levels below the pig's requirements can affect muscle development, and cause a fatter carcass. Dropping protein content in the ration will decrease feed efficiency. Slaughter hogs should be pushed harder than breeding animals, but only as long as they are producing frame and muscle. Excessive fat cover is the most expensive gain that can be put on a hog.

The difference between a pig in a poke, and pork on the plate is hard work and an eye for detail. And, according to the old-timers, to be really successful, you honestly have to like pigs.

PIGS IN THE PASTURE

We have discovered that raising two feeder pigs each year in our spare time is both an easy and economical venture. It also gives us plenty of lean, healthful, additive-free meat to put in our freezer each winter.

Since we have plenty of rough pasture land on our Pennsylvania mountaintop farm, we practice a three pasture rotation system. By trial and error, we have discovered that a pasture 40 by 100 feet is an ideal size for two pigs to thoroughly graze, root up, and fertilize before slaughtering time.

Having a clean pasture each year for our pigs, one that has not been used by hogs for two or three years, prevents disease and internal parasites, particularly roundworms. According to older government bulletins, good pastures can save one-third on protein feeds and one-fourth on grain. Forage crops are also rich in minerals and vitamins. In addition, this rotation eliminates any bad odors, since pigs are naturally clean animals.

The following year, that same pasture becomes an outstanding squash and corn garden. Despite no other fertilizer or manure and the annual bombardment of striped cucumber beetles, we harvested a bumper crop of summer and winter squash (zucchini, patty pan, spaghetti, acorn, butternut and buttercup), edible-seeded pumpkins, and Golden Cross Bantam corn last summer, from our former pig pasture.

After harvesting the squash and corn, we plant a pig pasture seed which grows through the entire third year before becoming a pig pasture once again the fourth year. The mixture we have tried is nearly equal parts of timothy, rye grass, alfalfa, and Kentucky bluegrass, but we were not entirely satisfied with it. Further reading in old government bulletins has convinced us that an alfalfa-bromegrass mixture would be the best seed to use for our farm.

Most of the time and expense involved in raising the pigs comes at the beginning, when it is necessary to purchase and erect fencing. Every year, for the first three years, my husband, Bruce, fenced in one pasture, which usually took a weekend of work. He cut locust posts from the edge of our field and used stock fencing to keep out curious pets, wandering chickens and ducks, and children. To pig-proof the fence, he strung a single wire of electric fencing 12 inches from the ground inside the stock fence. Twelve inches seems too high when the pigs first arrive, but we have tried lower wires and found that the pigs root out just in front of the wire, overturning clods of dirt onto it and shorting it out. Twelve inches is still low enough to give the pigs a couple of nose-to-wire experiences which teach them how far they can venture. We have never had to chase escaped pigs.

Once the fences are up, raising feeder pigs is easy. We like to buy six-to-eight-week-old pigs from a reliable pig farmer at the end of May. That way, when the cold weather arrives, in late November, they are ready to slaughter.

We cannot stress enough the importance of finding a clean operation, since pigs are very susceptible to disease. For that reason, we do not buy our pigs at auction. In fact, we found a really isolated pig farmer who never attends auctions or associates with other pig farmers. (We heard about him at our local feed mill, an invaluable source of all kinds of information, by the way.) When we called him up, he was most reluctant to sell any pigs, and we had to do some fast talking to persuade him that we were not pig farmers and would not bring any diseases with us. Since then, he only needs a little persuasion to part with two feeder pigs each spring at auction prices. Often, the pigs he selects are larger than feeder size, and they are always extremely healthy. We don't even have to use a wormer on them, usually advisable when pigs are first brought home.

Besides a pasture, pigs do need shelter from the sun and lots of water. My husband designed and built a simple, open, four-sided, slant-roof shelter out of scrap lumber he found in the barn. The shelter can be dragged from pasture to pasture by the tractor. When it starts to get cold in the fall, our boys fill it with hay, which the pigs arrange as warm bedding.

Since each pasture is near the barn, we have a hose hooked up, to keep two large watering containers full at all times. The pigs also enjoy a light shower bath on scorching hot afternoons.

In addition to the pasture, we also provide the pigs with three meals a day. Our eldest son, Steven, is in charge of the morning meal, which is usually a mixture of commercial pig feed and water, I supply the afternoon bucketful of garden leftovers, and my husband collects our edible garbage to mix with more feed and water at night. Pigs eat everything but orange, lemon, and banana peels, coffee grounds, and egg shells (which we recycle to our chickens anyway). They particularly love the innards, head, and feet of slaughtered fowl, and unhatched duck eggs. Other favorites are rotten tomatoes, fallen pears, and windfall apples. In fact, as the pigs grow bigger, we look on nearly everything as potential pig food. When we pull bean plants from the garden, down they go to the pigs. Also, they love old broccoli and cauliflower leaves, lettuce that has seeded, pea vines, and, of course, comfrey, which I grow especially for them.

Once they reach about 220 pounds, they are ready for slaughter. We call up our supermarket butcher friend, scrub down the basement of our barn, take out a couple old kitchen tables, and sharpen our knives. The pig (we usually slaughter them separately about three weeks apart) is shot between the eyes with a pistol, hauled up on a rope hoist, skinned and gutted. Then the meat is wrapped in plastic and chilled overnight. The following day, it is cut up, wrapped and frozen.

Our butcher friend so enjoys working with such high quality, lean meat that he usually does all the cutting up, and we do the labeling and wrapping for the freezer. For that, he takes some chops and a couple roasts and calls it even.

We do cure and smoke the bacon in a rather crude, recycled, outhouse-smokehouse my husband built, but we have neither the time nor energy to attend to the much longer process of ham-curing. So we freeze them and use them as we do the pork roasts. We feel it is better for our health anyway.

We also grind and season our own loose sausage and freeze it in one-pound packages. Since it is very lean, we substitute it for ground beef in spaghetti and chili recipes, as well as using it in standard sausage casseroles. Our bacon is also very lean, and it flavors our greens, replaces salt pork in baked beans, and accompanies fried pork liver. After smoking, I freeze the bacon, both sliced and in chunks, in small packages. We eat nearly everything and are particularly fond on the tongue and heart, which I cook overnight with water and spices in my slow cooker. Then I chill it and slice it for luncheon meat.

PIGS FOR PROFIT

Our family is probably like yours in one respect: working hard at farming for self-sufficiency, but holding down one outside job to pay the mortgage. Although the farm furnishes most of our food, fuel, and such, it's definitely not a get-rich-quick enterprise . . . it's more like a break-even enterprise, except for the pigs.

We bought our first brood sow as a piglet intended for the table. When we plunked down our $30 for a five-week-old piglet, no grand ideas about breeding pigs were in our heads; that came later, and I insist that there was nothing sentimental about it. She was just such a nice pig, with such a winning personality, that I thought, why not keep her and breed her?

Having determined that Blossom was ready (the vulva become a deep pink, every 21 days until she is bred), I called the man with the boar and then set about preparing the bridal chamber. Not knowing what to expect, I spent an entire day fortifying the pigpen with two by sixes and a pig-proof gate. All wasted effort. The groom, a perfect gentleman, went about the nuptial ceremony with the utmost delicacy and restraint. And in about four days, when the bride's attractiveness had begun to pall, he cracked the code of my pig-proof gate with the same gentlemanly aplomb. I glanced out the window one snowy afternoon to see my borrowed boar heading deliberately across the front lawn in the direction of Route Two and, presumably, greener pastures. Word of advice: no matter how pig-proof you think your gate, keep the barn door shut!

Breeding and Farrowing

You can begin, as we did, with a sow piglet (called a gilt until she has farrowed), or you can invest in a breeding-age gilt, five to seven months old and weighing 175 to 200 pounds. Unless you grow your own feed or have a source of free food, such as the school hot-lunch leftovers, the total investment is about the same. Although there are various breeds of pigs to choose from, I doubt that it is economically advantageous to raise purebred animals. The question becomes academic, in Maine at least, because little purebred stock is available. If my husband, who is an Irishman, had his way, we would raise nothing but Landrace, the long, lean, pink pig with the floppy ears. This is the only breed raised in Ireland, and it is so carefully monitored that any pig born that is not pure pink must be destroyed. At any rate, look for a long, lean pig (bacon type) rather than a short, fat one (lard type). Check to see that she has 12 to 14 working teats.

Gilts come into heat every 21 days, for a three-day period (sows will go three to five days). Like other livestock, they are in standing heat (will stand still for the male) only for a short time, say, 24 hours in the middle of the three-day period. This is important only if you are bringing your sow to the boar for "same-day service." Some owners of boars encourage this type of service, but most will welcome your sow for a few days, or rent you the boar for a few days, in which case the standing heat period is not critical. The fee is expressed in terms of dollars (usually $15) or a piglet, your choice. Unless you have five or more brood sows, it's not economically practical to keep your own boar.

The gestation period for hogs is three months, three weeks and three days (and sometimes, three in the morning). You will be feeding your sow four pounds of grain a day during gestation, with plenty of clean water and, if you're thoughtful, a small ration of bran during the final week of pregnancy. Five to seven days before she farrows, you will notice her "bagging up." Now is the time to give the pigpen a super clean-out, or bring your mother-to-be to a separate farrowing pen. Give her plenty of bedding material; she'll make herself a nest. Don't allow bedding to become too deep, though, or the piglets may burrow into it and be stepped on accidentally.

It is at this point that I part company with all the books on hog-raising, which recommend that you be present at the birth. I was on hand for Blossom's first farrowing, but only because it was at three in the afternoon. Every other blessed event took place late at night, and I was never one for sleeping with the pigs. In a rare instance, a piglet will get stuck in the birth canal, and you will have to roll up your sleeve and go in after it. Most of the time you'll find Mother Nature doing an expert job on her own, with no need for you to stand by. If you are present, expect the piglets to come 20 to 45 minutes apart in the first litter.

Now a piglet at birth is fully equipped for survival—more so than any other infant. It walks, it talks, it has eight needle-sharp teeth. The only thing it is susceptible to is stress. Left to its own devices, it will pick itself up, trot around to the dairy bar and go to work on breakfast. And I suggest you leave it at that. The books will tell you to dip its navel in iodine, clip its teeth, and set it under a heat lamp. I went through these ministrations once, and concluded that the dubious benefits of all this activity were offset by the stress on the piglet. My advice: let Nature do it her way. She's had a lot more experience in these matters than you or I.

Regarding the great Afterbirth Debate: the same people who recommend fiddling with the piglets also urge you to snatch that afterbirth away lest the sow eat it. They claim it might give her the idea that the piglets are edible too. As you can guess, I side with the Mother Nature school again. Sows have been eating afterbirths for evolutionary eons before man interfered, and the race of pigs survived. Besides, I have a hunch that all that rich protein may be just what the doctor ordered. And anyway, you will probably walk into the barn on the morning of Target Day to find Ma grunting contentedly as 10 little fellas belly up for brunch, and the whole matter will be out of your hands.

Immediately give the new mother plenty of water, not ice-cold. During the next three days, in addition to a constant supply of water, gradually increase her feed ration until you have doubled it— eight pounds, four in the morning and four at night. Your biggest worry in these first few days is that the mother will step or lay on her babies. One precaution you can take is to keep your brood sows on the lean side, decreasing the odds that they'll squash a baby without noticing. Another precaution is to nail two by four or two by six guard rails, edge out, about eight inches above the floor on all sides of the pen. This prevents piglets from being squashed between mother and the wall or floor when she lies down to nurse.

The next milestone in the rearing of piglets is injection of iron and castration of the males, performed when the piglets are five to 10 days old.

Iron Shots

Baby pigs are born anemic and need an iron supplement, the textbooks say, to thrive. This can be a liquid painted on the sow's udder at intervals, or a one-time injection of Iron Dextran. I prefer the latter, since I can treat the whole litter in a few minutes and be done with it. If you've never given an injection before, this is as good a way as any to learn. Arm yourself with a small syringe and a small disposable needle (one needle will do a litter). Disassemble the syringe and throw it and the needle into boiling water for 15 minutes. Meanwhile, gather the materials you will need for castration, since you will perform both operations while you have the litter in the kitchen. You will want a razor blade (single edge is best), antiseptic, bowl, clean rag, and an assistant. (Note: if you are truly squeamish about all this, you can hire a vet to do it. However, if you're in the pig business to help pay the mortgage, there's not much logic to putting your profits in the vet's pocket.)

The iron shot is injected intra-muscularly (IM) into the neck muscle or ham. I prefer the ham, since there's less margin for error, and if done early, no staining of the ham. Holding the syringe needle-up, and the bottle of Dextran upside down, plunge the needle into the bottle and draw off one cc of fluid. Remove the bottle and press gently on the plunger to dislodge any air bubbles. While your assistant holds the piglet steady, jab the needle into the ham a bit more than skin-deep, and slowly empty the syringe. Presto! That's all there is to it.

I mentioned earlier that the books claim the iron is necessary. I have raised pigs for the table both with and without iron supplements, and have seen no apparent difference in rate of growth. But I continue to administer it on trust because (1) the experts say so, (2) it's cheap and easy, and (3) some customers expect it.

Castration

An uncastrated boar is not edible pork. Boar piglets must be castrated, except when saving one for breeding. The ideal age is five to 10 days; at this age the operation is easy, practically pain-less, and nearly bloodless. Have your assistant, sitting opposite you, hold the piglet upside down by the hind legs, its shoulders between his knees, its belly out. With your blade, make a vertical incision about an inch long between the two testicles. Make sure that the white membrane surrounding each testicle is also severed. Work one testicle out with your fingers and give it a quick snap. The cord that holds it should break (if not, cut it). Drop it in the bowl. Re-peat with the other testicle. Swab or spray the incision generously with antiseptic. The rag is for catching drips.

In my experience, there are always more females than males in a litter, so doing the boars takes only a few minutes . . . it goes faster with a bit of practice. Do keep an eye on them for a day or two to make sure no infection sets in (none ever has, for me). You may dispose of the by-products from this operation in any way you wish; I feed them back to the sow on the theory that extra protein is good for her, but this practice may be too cannibalistic for your liking.

A word here about a minor point the books always fail to men-tion: how to catch the piglets at castrating time and, later, on "mar-ket day." Sows are very protective, piglets have an ear-piercing squeal, and your heretofore gentle mother may just take your leg off if she becomes alarmed at the high-pitched noises of pig cap-ture. The only really safe way to proceed is to separate the sow from the piglets. At times, I have lured her into an adjacent pen with corn. Usually, though, I have a portion of the pen walled off with two by four's so that the piglets can creep under to be fed, but mother cannot follow. This creates a demilitarized zone where piglets can be caught (by a hind leg) without risk to life or limb.

Feeding and Weaning

At about 10 days to two weeks of age, your piglets should be ready for solid food. Offer them a pan or shallow trough of grain mixed with water or surplus milk to gruel consistency. At first they will waste more than they consume, but soon they'll be looking for-ward to their feeding with typical piggy anticipation. Once they be-gin to eat, you can mix their feed to a thicker consistency . . . then stand back!

At five weeks, the litter is ready to be weaned and sold. Drop the sow's rations back to four pounds of grain daily. In 18 to 24 days she will come into heat and can be bred again. You will have both a spring and fall litter to sell, since sows are bred twice a year. In all likelihood, every one of your piglets will have been spoken for well in advance of the weaning date. At first, you may have to put up a notice in a local store or market, but once the word is out that you have pigs, people will be calling you. As a last resort, you can take your piglets to auction. You will get a good price, but will have to deduct your time, gas and the auctioneer's commission. For several years the price of spring pigs held steady at $30, fall pigs at $10 to $12. This spring, the price went from $35 to $45 for spring pigs, and $15 to $20 for fall pigs.

Let's take an overall look at the economics of breeding pigs. It cost us about $115 in purchase price, feed bill and boar fee to bring our first sow piglet to spring weaning date. The first litter, of which we sold eight, brought in $242, and we kept the ninth piglet for the table. Total profit: $157, including our freezer pig. The fall litter that year brought in $102.50 (10 piglets) plus a roast suckling pig for the table—a profit of $3 over and above the grain bill and boar fee. The following spring, 10 piglets brought in $300 (plus two for our table), netting a profit of $220, plus our "free" pork. In short, the fall litter pays the feed bill, the spring litter is the moneymaker. Add more sows, as we are doing this fall, make more money.

Since we modified existing structures with scrap lumber, our housing/fencing costs were zero. However, in addition to the boar fee, we also have to pay someone to truck the boar to the sow, or vice versa. The value of the piglets we raise for the table must also be figured in. Considering that our freezer pigs are "free," and adding up the cost of grain, butchering and smoking, the average cost of our ham, pork, bacon, and lard, is 40 cents per pound. With the price of bacon alone at $1.79 per pound, you can see that not all the profit in pigs comes from the sale of the piglets!

Chapter 6

Cattle

With the availability of types of cattle today, the following random sampling should provide a reasonable spectrum for consideration. Whether a cow is to be used to work the land, provide milk, or stock the freezer the articles provided here should help to make an informed decision.

A COW-BUYERS GUIDE

Many are the pitfalls that await the tenderfoot back-to-the-lander who first attempts to stock the new farm with productive four-footed beasties, whether from a stockyard or directly from a farmer. The following dissertation should help you to avoid making the more common mistakes a greenhorn makes when buying the first family cow.

Calves You Plan To Raise to Butcher

Look at the Nose. Is it running like a faucet? If so, the animal may have a chronic cold at the very least. Is the nose dry? The creature probably is running a fever, indicative of pneumonia. The muzzle should be moist, but not dripping down to the knees.

Watch the Feet. There's a good possibility that a limping animal has foot rot. This is a contagious, hard-to-cure, probably chronic disease. The foot rot bacteria is next to impossible to eradicate from your ground, once established. Don't purchase a gimpy animal.

Check the Hair. It should be smooth, shiny and soft. If patches of it are falling out, the little dogie could have lice, mange, scabies or any number of other ailments.

Examine the Eyes. Watery, running eyes promise to progress to pinkeye in the near future and a critter with such eyes should be avoided.

Watch the Price. It's best to pay market price for any purchase. If no one will bid on an animal at an auction, there's probably a good reason why. Ask the wise old farmer sitting next to you why the beast is selling so cheaply.

The best money you'll ever spend will be the small expense of having a veterinarian look your newly acquired critter over before taking it home after the sale. If something is wrong, or if preventive shots are required, he'll take care of it for you.

Baby Calves

Never buy a baby calf unless you know for sure that the youngster has had some of its mother's first milk, known as colostrum, or unless you have some colostrum on hand. A shocking number of baby calves sold both from the farm and at auction, have never tasted a drop of their mother's milk. I cannot stress enough the importance of this one point. Find a neighbor, friend, or relative who has a cow that has just had or is about to have a new little one, and ask them to save you the first milk they get from the beast. Goat's milk from a newly fresh nanny will work just as well as cow's milk, or better. Freeze the milk in one-cup or one-pint-sized containers and you'll always have some available when needed.

Look at Their Tails. If a newly hatched calf has white manure all over its tail and behind, it has infectious scours (diarrhea that spreads from animal to animal) and should not be touched with a 10 foot pole. This form of scours is contracted through the navel, and once it gets its bacteria into your soil you can expect to spend years fighting the disease. Infectious scours can wipe out a whole herd of calves and keep on doing it year after year.

Check Their Navels. A baby calf with a wet navel is sure never to have had any of its mama's first milk. I hesitate to buy one whose navel hasn't dropped off entirely unless I have colostrum to start it off with. Newborns' tummies are very touchy.

Many times when a group of stock cows (beef-type range cattle) with calves at their side are sold, the yardman will split up the pairs. These calves will definitely have had a good start and would be your safest purchase.

Stock Cows

A stock cow is a beef-type animal that you plan to keep for the purpose of breeding and raising calves. Most stockyards now pregnancy-test all beef cows brought in. When the auctioneer says something like, "All pregged and ready to go," the cow will eventually have a little future sirloin. Always insist on a pregnancy test when purchasing stock cows that are being presented to you as bred (with calf) whether at a sale or from a private party. Look potential baby-producers over for lumps on the jaw or brisket, rough hair, and spoiled udders. A spoiled udder, or bag, will have a hard, lumpy quarter and will sometimes have pus dripping out of a hole or a teat. Don't buy such an animal. Keep your eyes open for pinkeye signs. Most states require Bangs (also called brucellosis or contagious abortion) and tuberculosis tests before animals can be transported. If there are not such requirements, ask that the tests be conducted on any adult cows that you may buy. Don't buy a cow who carries her head high, and acts very nervous and "snorty." Such a bovine will probably take off for the high country, fences or no fences, the minute you let her out of the truck. It's always a good idea to put a strange beast in a good, tight corral or barn for a couple of days after hauling it home, to allow its nerves to settle down before turning it into a larger field.

Lactating or "Milk" Cows

An amateur should never buy a milk cow through an auction. It's just too hazardous. Watch your newspaper's classified ads section for a cow and one will turn up sooner than later. A little patience here can save you a lot of money in the long run.

When you start out to look at a potential milk cow, take mastitis test cards with you. These can be purchased anywhere that veterinarian supplies are sold. Instructions for the use of the cards are on the box they come in. If the cards test even suspicious, turn the animal down. Mastitis is a condition in which the cow's udder becomes feverish and sometimes hard. Her milk becomes stringy and lumpy. The disease can be cured, but not, usually permanently.

Never buy a cow that has not yet calved (given birth) unless the seller will guarantee her to be sound. Get it in writing! Insist upon a brucellosis (Bangs) test and a tuberculosis test on any milking bovine you may wish to purchase. As with a stock cow, if the milker has not yet freshened, ask for a pregnancy test. Before buying an animal that is already lactating, insist that you be allowed to milk her at least twice, at a 12 hour interval. This is the only way to find out how much moo-juice the bovine gives, how "easy" or "hard" milking she is, and how tractable she is. While milking Bossy, feel her udder carefully for lumps or sores between the quarters. Udder sores are usually a sign of a serious condition, and a cow with such sores should be left strictly alone.

How Old Are They?

The only relatively sure way of telling age on an animal (and please note, I said relatively sure) is by examining the teeth, or mouthing it. A young creature has all his teeth and they stand straight up and even. As the brute ages, the teeth begin to spread, or to splay out. As the beast's years further advance, it begins to lose a few incisors. Simply, if the animal has a mouthful of pretty good, healthy-looking choppers, you'll get several good, productive years out of it.

General Good Advice

Try to buy good, healthy-looking animals rather than bargains, and have a veterinarian look them over for you after a purchase, or before buying whenever possible. Look out for lumps, bumps, sores, rough hide, running eyes or noses, messy behinds and lameness.

Before attempting to buy at auction, attend a few sales just as an observer. Learn to understand the auctioneer and, most important, learn to tell the difference between an animal being sold by the head and one being auctioned off by the pound. When an animal is selling by the head, the price goes up at least 50¢ to $1.00 at a time. When the sale is by the hundredweight (or by the pound) the price goes up 10¢ at a time. More than one novice buyer has purchased a dogie at what he thought was, for instance, $45. The auctioneer had concluded the sale with the statement, "Sold out to So-and-So, $45!" The tenderfoot thought the price for $45 by the head (for the whole animal), but the auctioneer actually meant $45 per hundred pounds. For a 300 pound animal, the actual price, then, was three times $45, or $135. Learn to differentiate before you bid.

Always trust your eyes, not what the seller tells you about the critter he's trying to sell you. Something seems to come over even an honest person when he undertakes to hawk his goods, and he exaggerates all over the place. Never was the phrase "Caveat Emptor" more appropriate than during a livestock transaction. Insofar as livestock auctions themselves are concerned, there's one cardinal rule you must remember. "What you see is what you get." If you buy a nice, healthy steer (a castrated bull) through a sales yard and the day after you get it home, it turns out to be a bull with lice and pneumonia, don't bother complaining to the stockyards, because no one is going to listen to you. The theory is simply that you shouldn't have been buying at auction in the first place if you didn't know enough to differentiate between a sick bull calf and a healthy steer.

Before you nod your head, scratch your nose, wriggle your fingers, wink your eye, or otherwise indicate your desire to bid at auction, be darned sure the quadruped being paraded before you is really what you want to buy.

YOU WANT TO OWN A COW?

If you are thinking about buying a cow, there are several things you should consider. First, you should be sure that you like cows, for you aren't going to own a cow—she is going to own you. A cow knows a lot more than she is given credit for, especially if you bring her up from a calf. She is a creature of habit and responds to regular attention. Cows are as individual as people. Some are gentle, affectionate, and calm. Others are nervous, high strung, and unpredictable.

Secondly, while your cow is giving milk, she is a seven-day-a-week job. During her dry period, if she's on pasture, you can have a rest, but you still must be sure to check her every day. Cows are valuable now and cattle rustling has come to the Northeast.

Third, consider how much land you have. A cow doesn't require a lot of it; if you own an acre or two that you can fence, she will be able to get her own roughage during the spring and summer. The amount of land needed to pasture a cow depends on what condition the land is in; i.e., whether or not it has a good hay and clover cover, or just weeds and witchgrass. If you don't own an acre but have a backyard big enough for her to exercise in, you can still keep a cow if you aren't in city limits governed by zoning laws. Baled hay doesn't take up much space and if you buy 10 bales or so at a time, you can house her in a smaller building than you might think. Be sure she has room to move around and a good, dry floor under her. You could house two Jerseys in an eight by 16 foot building, with room for up to 20 or so bales of hay.

There is an old saying that a cow is "half your living." It's true, you get a lot from a cow: not only milk, but cream, butter, cheese, fertilizer and grass cutting. With milk at nearly $1.82 per gallon (and this just milk with hardly any cream) you can figure a cow will give you more than four dollars worth of milk a day.

If you raise a beef animal, you also get a lot of meat. A beef animal will dress out at an average of 55 percent of liveweight. So a thousand pound animal will give you over 500 pounds of beef, and probably cost you between two and three hundred dollars in all. (And a beef animal is much less work, as you don't have to milk her every day.)

You should plan to keep a calf eight weeks for veal, and during this time, it should nurse its mother. If the calf only nurses, the veal will be white, but if it's fed grain and hay, the meat will have more color and be less tender. Yearling beef is often pale and stringy; neither veal nor beef. For beef, keep the animal two years. In the last two months before butchering, confine it, and gradually increase the grain consumption to make the beef tender.

I have raised 300 calves over a 42-year period. I had a cow named Cocoa for 13 years. She was a big white Holstein with black ears, who gave a lot of milk and gave it for a long period. She was the Boss Cow. In a herd of cows, there is always a Boss Cow—she gets the food first, she leads the herd to pasture, and back to the barn, and expects to be milked first. Cocoa was one of the only really jealous cows I ever saw. If I should step into the barn and start to pat another cow before I spoke to her, she would start hooking the cows next to her to evidence her displeasure until I paid her all my attention. With me, she was gentle as a lamb, but very mean to the other cattle.

My mother had the smartest cow I ever saw; a little mixed breed named Chee Chee. Mama raised her from a calf and Chee Chee was perfectly happy as long as she could see Mama. Dad bought two more cows and they were all in the pasture behind the house where Chee Chee could see Mama in the kitchen window. One day when Mama was washing dishes, Chee Chee came running, looking in the window, and bellowing as loudly as she could. Mama went out to see what was wrong and found the other two cows out of the pasture. After that, the other cows got out on several occasions, but Chee Chee always turned them in.

After you've weighed the pro's and con's and have decided to get your cow, you've come to the next step: Do you plan to raise her from a calf, or buy a cow ready to milk? Unless you buy the grown cow from a person you know and trust, it's better to buy a first-calf-heifer just before she freshens (calves), than an older cow. The older cow could have problems. She may be a hard milker, not have much cream in her milk, have chronic mastitis, or be mean to handle. A first-calf-heifer, never having been milked before, can have problems too, but you won't be getting someone else's mistakes. To step out today and buy a full-grown cow costs quite a bit of money and is often beyond the reach of the average family, as their prices range between $300 and $500 and up.

On the other hand, you could start with a calf or two, which range from $20 to $50 or so, depending on the season. Calves are usually least expensive in late June and July, because this is the time they are most plentiful. Two calves do better than one because they keep each other company in summer, and warm in winter. When you only have one cow, you'll be without milk for 10 to 12 weeks of the year—you have to let her dry up her milk and rest for a couple of months before her calf is born. But with two, you can breed them so that one calves in the spring and the other in the fall, so you'll always have milk.

Late summer is the best time to start a calf, so she can get some age on here before our cold Northeastern winters set in.

When buying a calf, however, remember that it will be all of two years before she freshens and has any milk for you. On the other hand, her cost will be spread over two years and probably total a couple of hundred dollars. If you have the room, you could raise three and sell one when she calves, to pay for the others.

Before you buy, you also have to think about what breed of cow you want. There are several breeds raised for milk, and some are dual-purpose breeds like the Brown Swiss or milking Shorthorn. But for a family interested mainly in dairy products, you can't beat a Jersey cow, although Guernseys are good, and Holsteins as a rule give lots of milk and little cream. If it's milk you want, the Holstein is the best buy, as she is worth more than a Jersey for meat or sale when you get around to selling her. But if you plan to make your own butter and want real clotted cream (so heavy you can hang it on a nail) buy a Jersey. But don't forget the mixed breed or grade cow who may (or may not) retain the best qualities of each breed and give you lots of both milk and cream. Most farmers breed their cows artificially now and this makes the average cow much better quality than she used to be, when any old cull bull was used for breeding.

If you decide to buy a calf, you can do a lot to keep her healthy and growing by following a few simple rules.

If you have two calves, keep them separated so they can't suck on each other, because they can ruin their udders doing this as babies. They will also chew on a cow or horse's tail, or try to eat old burlap bags or papers. Never leave a calf (or cow) where she can lap on a painted surface, in case the paint contains any lead.

Never hitch a calf without a swivel in the chain. Or, you can snap a swivel snap into her collar. This prevents her from getting her rope or chain twisted up and possibly hanging herself.

Never buy a calf that hasn't had the colostrum from here mother. This is the most important thing in giving the calf a good start in life. Colostrum is a thickish yellow fluid that the cow has in her udder when she calves, and it contains nutrients that are essential to the calf's well-being. For the first 72 hours, the calf should either nurse the mother, or the mother should be milked and the calf fed the colostrum. Within the three day period, the colostrum changes to milk and then it's okay to take the calf from her mother and feed her milk or milk replacer.

Most major feed companies manufacture a product to take the place of milk. This usually contains powdered milk and vitamins, and most have an antibiotic added to help prevent disease.

Never over-feed your calf. Feed her three times a day for the first month, about two quarts to a feed, according to the calf's size and directions on the product you are using. Have food warm—a little more than lukewarm, but far from hot. Feed at regular intervals and in a clean container. Offer her a chance to drink a little water (with chill taken off). Keep her bed dry with old hay or straw, but do not bed a young calf with sawdust or shavings, as she may eat them and get digestive troubles. Keep her out of the hot sun in the summer, and never leave her in a draft or in a dark, damp area.

After the first week, offer a handful of good quality hay and get her to lap a little coarse Fitting Ration (grain) from your hand. She will spit at first, but soon like the taste.

Coarse fitting ration is what I recommend feeding the family cow—this is a little over $8.00 a bag at present and should last one cow at least two weeks. You can pick up a lot of roughage for your cow from neighbor's cornstalks, beet tops, and lawn trimmings. Hay is now bringing a dollar a bale; a bale of hay will probably last two days, depending on its size and that of the cow. Do not feed her any of the cabbage family.

As near as I can figure, it will cost you around a dollar and a half a day to feed your cow, and she should return at least double that, not counting your labor (count that as exercise good for you).

Besides hay or grass, and grain; cows need water and mineralized salt. Water them at least twice a day; usually three or four pails a day is all they want. The easiest way to give them salt is to buy a salt brick; a colored one that contains the trace minerals they need to prosper. Put it in their crib where they can lap it when they need it, or put one in the pasture.

An easy way to water them is to find an old bathtub (check local secondhand stores or even the dump) and run your garden hose into it. Fill it once a day in the pasture and it saves a lot of work if you aren't fortunate enough to have a brook or pond they have access to.

There are several plants that are dangerous to a cow or calf, the most common being several kinds of ferns, Indian Polk (skunk cabbage), lamb kill or forms of mountain laurel, rhododendron, oleanders and azaleas. Castor Bean plants are poisonous also—this is a list of only a few of them. Watch out for any kind of cherry tree that has wilted leaves. The leaves are only dangerous when wilted; when green or dry, they are harmless. During the wilting, an acid is formed that is deadly to livestock.

You will avoid a lot of trouble with poison plants if your cow is supplied with a lot of good feed and not turned out too early in the spring when grass hasn't sprouted.

Outside of poisonous plants, there aren't too many serious diseases to worry about except mastitis. This is caused mainly by unsanitary conditions and irregular milking. Her udder will feel hot and hard to the touch and in acute cases, the cow will go off feed. It could be dangerous. You can buy tubes of antibiotic udder infusion at the feed store, but in an acute case, it's best to get the vet.

If your cow's udder ever feels dry or cracked, rub it liberally with Bag Balm (available at feed stores). Never leave a quarter (teat) unmilked because it gets cut or injured. If the canal is damaged, you have to open it. If it isn't serious, you can buy medicated teat dilators that you insert into the injured teat to keep it open. But if the injury is very bad, it's best to call a veterinarian.

Some really good cows will develop milk fever after calving. She will not be able to get on her feet. Keep her from rolling over on her side by propping her with a bale of hay, while you wait for the vet to give her a shot of calcium chloride or something to give her the needed calcium. Fast treatment is required.

Brucellosis or Bangs Disease is very uncommon, at least in

Maine right now. In fact, Maine is listed as Bangs-free at the moment. With good regular care, you should avoid most veterinary bills.

Strange as it sounds, there is much less danger of a cow getting pneumonia or shipping fever if you move her from a warm area to a cold area, even outside, although I believe Jerseys need to be housed. To bring a cow from the outside cold into a warm barn sometimes results in her getting sick.

Yearlings and beef cattle winter nicely if they are allowed to run loose, with a shelter from the worst cold winds and storms. A barn cellar with free access to the yard or pasture all winter is ideal for them. Just be sure they have water.

In summer, horse flies and face flies can cause cows to gain poorly, because they keep them on edge. There are good dairy fly sprays at your feed store.

In winter, cattle can get lice, especially if there is a flock of poultry nearby, and the constant biting will give them an unthrifty appearance. If you check carefully, you can generally see the lice. A good louse powder will take care of the problem easily.

I should mention the danger of bloat in a cow. This is usually caused by the ingestion of too many green wet legumes, such as clover or alfalfa. It can be prevented by feeding your cow fodder or dry hay before turning her out, so that she won't want to eat again until the clover has dried off. Bloat can also be caused by moldy hay or grain. If your cow gets bloat, never let her lie on her side.

Bloat can be dangerous. If veterinary help is not too far away and the bloat is really bad, the vet can relieve it with a tro-car, a little sharp instrument he uses to punch a hole in her stomach to release the gas.

It would probably pay you to buy a book on the care of cattle, that goes into detail on every problem a cow or calf could have.

The Wholesale Veterinary Supply Co., P.O. Box 2256, Rockford, Ill. 61131, has a large animal catalog that lists every kind of instrument and medicine you would ever need, at reasonable prices.

Don't try to handle a cow or calf with a rope around her neck. Make a nose halter from a length of rope by making a noose that fits the animal's nose. Bring the rope over the animal's head behind the ears and horns, then bring it down on the other side of the head, and tie it into a noose. You can control a large animal with a good halter—you can buy one ready-made.

Tie chains are the least expensive hitch for the cow. That's all I've ever used. Stanchions are more expensive by far, but not necessary for a family cow or two. I have hitching poles made of hornbeam (a native Maine wood) that is practically unbreakable. The tie chains slip up and down on the pole, letting the cow lie down or eat in comfort. On the bottom, use a six by eight inch or large beam; same on top. Bore holes through the top and bottom beams to fit your hitching poles in. The bottom timber also serves to keep her feed out from under her feet so she won't waste it.

An hour a day, including the milking, should be ample time to care for two cows. With one or two head, it's easier to milk by hand than use a milk machine, as it's more work to keep the milker clean than it is to just milk by hand. Milking is simple if you understand it. When you grasp the cow's teat in your hand, you squeeze, but you move your thumb and forefinger gradually down the udder shutting your hand into a firm, gentle squeeze. But first, you should wipe your cow's udder with a warm, wet paper towel that's been dipped into hot water with a drop of Clorox. This prepares her for milking, because it stimulates the hormone that lets down her milk. Once you start milking, don't stop until you have milked every teat dry. It might be helpful to visit a nearby farmer at milking time and ask him to show you the proper way to milk.

If you have raised your calves yourself, you will want to plan to breed them at about 15 months of age, so they will calve at approximately two years old.

Dairymen with Holsteins and other large-size cows sometimes let their heifers get close to three years old before calving, but for the little family cow, two years is okay, if you have raised her well and she is in good condition. She will continue to grow even after calving. You can use artificial breeding. Check with the nearest farmer for your local inseminator, or maybe he will have a good bull you can breed to.

A cow comes in heat approximately every 18 to 21 days, and it usually lasts about two days. It is best to breed the second day of heat. You can tell a cow is in heat by her behavior. She will be nervous, bellow, and have a colorless discharge, followed, in two or three days by a little show of blood. She will try to mount any other animal—even you—so beware!

Years ago, I watched my father-in-law do a strange thing. Immediately after the cow was bred, he stepped alongside her and doused her with a couple quarts of ice water. I thought he was crazy. He said he'd noticed that the cows were more likely to "stick" with calf if you did that, but he didn't know why. Later I was telling our veterinarian about it, and he said the reason was simple—the cold water shocked her system and caused her to drop her egg. So you see, some of the old ideas have a scientific explanation after all.

Watch your cow after breeding. If she didn't stick with calf, she will come in heat again in about three weeks.

Cows carry their calves about the same time as a woman, around nine months. Different breeds may go a few days one way or the other. Most of them will calve without trouble unless they have been bred to a big-boned or big-headed bull like a Holstein or Hereford.

Always keep watch of your cow when her time is out. She may need a little help, but don't bother her if things are going okay. Usually a pinkish water sac will appear first, then when this ruptures, you should see the calf's two front feet and nose coming on top of them, a short time later. If you see only one foot or the tail first, you will need help. Unless you have helped and know what to do, you should call the vet.

If both front feet are there, give her time, but if she continues to strain with no progress, a little pull may be the answer. Wash your hands thoroughly with soap and Clorox in hot water. Fashion a slip knot of baler twine or other strong cord around one front leg, then around the other. Make sure the nose is in position, and the head not bent off to one side. Tie the other end of the cord to a short, strong stick. Grasp the stick in both hands, and stand so that when you pull, you will be pulling down toward the cow's hind feet. When she stains, start a steady pull. When she stops that stain, hold what you have gained, but don't pull until she strains again. However, once the calf's head and shoulders are out, keep him coming so he won't stick at the hips. He will break his navel cord himself. If his nose is covered with membrane, wipe his nostrils and mouth clear so he can breathe. Usually, the mother will jump to her feet. Be sure she can reach her calf to wash him with her tongue. The rough tongue starts his circulation, and gets him going.

Shortly, the calf will get to his feet after several attempts and try to nurse. You must be sure he nurses. If he can't seem to do it after a while, milk out a little of the colostrum and carefully turn a little of it into his mouth from a small dipper. Leave his lower jaw free so he can swallow. The only danger to turning any liquid in a calf or cow is if they can't swallow. Don't turn it in too fast. You may have to steer a teat into his mouth to start, but usually the mother and calf will get along okay by themselves.

A short time after calving, the cow will lie down and expel the afterbirth, a sac with bloody bunches that the calf was carried in. Mixing some fitting ration with a half cup of sugar and hot water will help her. But mainly keep things quiet while she does this and keep people away from her.

Some heifers will have an awful time over their calf—rolling him around, bellowing, and carrying on like mad. A few won't own the calf, but these are rare cases.

Don't let the cow be milked out dry for at least 48 hours. The third day, milk out anything left in her udder after the calf sucks. If you plan to use the milk, separate the calf from the mother on the third day and start feeding him yourself, as I instructed previously. She will try to hold up her milk for a day or two but soon will let you have it. There will be a lot less bellowing if you hitch the calf where she can see him, but he can't reach to nurse.

We average about five quarts of milk twice a day from our little Jersey heifers, but you can expect more each time they calve, for up to four years.

Several layers of cheesecloth will do to strain your milk immediately after milking. Then place it in the refrigerator to cool. If you plan to make butter, strain your milk into open pans so the cream can rise and you can skim it off easily in about 48 hours. Otherwise, for milk, pour into jars.

You should be able to eat all the cream you want on your berries and have your own fresh butter. If you are getting plenty of exercise, don't worry about dairy products hurting you unless you have been warned by your doctor. I truly believe people don't get half enough hard physical labor now. That's why people didn't worry about what they ate in the old days, only about getting enough to eat. Hard work was the answer.

Now, do you still think you want a cow? I wouldn't be without one, I have raised more than 300, and they are indeed the farm family's friend.

A PAIR OF OXEN: THE BEAUTY OF THE BEASTS

Roland Boardman is a lot like many New Englanders. He has a family, a regular job and enjoys working around his home during his free time. What sets him apart from most, however, is an unusual hobby that is not only enjoyable but helps him to save a substantial amount of money. He owns a pair of oxen.

"I've always been interested in animals," says Boardman as

he proudly turns and gestures toward his oxen. "I've loved all kinds of them since I've been a boy."

In the past, oxen were used to perform valuable services for man. The ancient Greeks, Hindus, Romans and Egyptians transported goods, dug irrigation ditches and plowed their farm lands by yoking and training the powerful oxen. They came to believe that there was a mystical relationship between these animals and the successful growth of their crops. In fact, the oxen were held in such high esteem that ancient law provided for their protection by stating that if a man should steal an ox, he shall restore it with five. Oxen were symbolized in drawings and statues and worshipped as gods. The ox was first domesticated when he emerged from the ark, working alongside man for centuries and an important factor in the advancements of some early civilizations. The Greek poet Hesiod, when referring to their affinity for work and handling put it this way:

"For draught and yoking together, nineteen-year-old oxen are best, because, being past the mischievous and frolicsome age, they are not likely to break the pole and leave the plowing in the middle." Boardman is taking advantage of their historically proven capacity for work.

Today in the United States oxen are still used to perform practical functions for man. Their use, however, is largely restricted to the southern Appalachian and New England regions. Roland Boardman is one of the relatively few Massachusetts residents who have raised, trained and used a team of oxen in place of modern day machines.

The acquisition of Star and Bright seemed like the natural thing for Boardman to do. The owning of several dogs, a pair of cows and a horse helped to nurture his present interest. A visit to the Anheuser Busch plant in Merrimack, New Hampshire, however, meant the difference between a longtime dream and the actual purchase of his pair. Draft horses and some oxen are kept at the plant, and when Boardman saw them, his mind was made up. In May of 1975, he bought his first ox from a local dairy farmer for $10. Less than three months later he obtained the second of the pair of black and white Holsteins. "They were a week old and weighed 100 pounds when I got them," say Boardman. "I've had as much enjoyment raising and training them as I have from owning any other animal."

Star and Bright stand outside a small red barn beside their master. They snort, and toss their huge heads back and forth to keep bothersome flies from settling on their backs. Dust is raised from their hooves as they scrape the ground, impatiently waiting for the next command. I wonder what terrible damage they might do if they turned upon us or the yoke was removed from their necks. "Don't worry," says Boardman. "They're perfectly harmless if you know what you're doing."

"Haw!" Boardman shouts and the oxen turn to the left, gently prodded by a whip with a lash that Boardman holds in one hand. "Come up," he commands, and his pair move smoothly forward in the direction of the barn that stands roughly 40 yards behind the family home.

The oxen move about as fast as a man can walk. Roland Boardman moves beside them and calmly guides them in the proper direction. As they approach the heavy wooden door to the barn he confidently orders them to stop. Boardman releases the yoke from their necks and directs them into separate stalls inside the barn. Star and Bright immediately dip their heads toward buckets of water. It is hot and they have pulled a plow for two hours now—hard work for both man and beast.

Roland Boardman stands relaxed beside his animals. He is wearing work boots, blue dungarees and a shirt with the sleeves rolled back. Suspenders and a hat top off the outfit. He is perspiring freely and removes his hat, one elbow is propped against a stall. He uses this hat to wipe the seat from his brow.

While it is simple to define an ox as a castrated bull, the difference between an ox and a steer is sometimes misunderstood. A steer is used for meat while an ox is used basically for work. It is not uncommon, however, for a lame ox to be killed and eaten should the animal be unable to perform any useful functions. The life expectancy of a normal ox is roughly 15-20 years.

"They are much easier to train than a horse, and they can actually do a lot more," says Boardman. "From my experience I'd say they're calmer than a horse and less stubborn than a mule. They might kick now and then, but generally, they are quite docile."

Boardman has trained his animals to readily respond to voice commands. To order the oxen forward the words "come up" are used. To order them backward the command "back" is given. "Gee" is used to turn the oxen to the right and "haw" is used to go to the left. When he calls out the word "whoa" his team comes to a halt.

A goad stick, or a whip with a lash, is used to guide the oxen in the proper direction when the voice commands are given. Boardman exercises Star and Bright daily, walking beside them voicing his commands and prodding them with the whip whenever necessary. He practices with them for about an hour each day.

Star and Bright may not be worth their actual weight in gold, but Boardman has found their services to be profitable as well as interesting and fun. "I guess I'm a lot like a teenager with his first car," Boardman admits. "The novelty of taking care of these animals and have them work for me hasn't worn off yet."

During the winter months, the six member Boardman family home is heated solely by the burning of hardwood. Star and Bright annually haul several cords of wood from his lot on a cart that Boardman built himself. "When you go out in below zero weather, a tractor might start, but then again, it might not," says Boardman, with the hint of a smile on his face. "I don't have that problem with these animals." Indeed, Star and Bright have been in places where a tractor or a horse just won't go. During the cold months, four or five feet of snow doesn't stop them from towing the wood their master has cut. And in the spring, thick mud makes parts of Boardman's land inaccessible to modern day machines. Star and Bright simply drive right through it. "An ox has a split hoof which gives it a greater pulling power," Boardman explains. "And they can be shoed when the ground is rocky or covered with ice."

Oxen are usually harnessed by a yoke that is fitted to their necks, but there are basically two different types of yokes that can be used. Neck yokes, the most popular, are made from wood that is cut, fitted and bent during the winter so that their shape will be maintained. This yoke is fastened around the necks of the oxen and their horns prevent it from slipping off. The less commonly used head yoke has a similar purpose but is fastened around the horns of the team. Star and Bright wear a neck yoke that was purchased at a flea market for $15. "If you have the talent," says Boardman, "you can make the yoke yourself."

If one of the pair tends to work harder than the other, a special, compensatory type yoke can be used to properly balance the team. This yoke, known as a sliding yoke, can be manually adjusted to equalize the pull. "Bright is a little lazy at times and would prefer that Star do most of the work," Boardman says. "The sliding yoke helps to take care of that problem." Boardman points out, however, that aside from the one he owns he has seen only one other. This yoke was found in the Heritage Plantation in Sandwich, Massachusetts.

In addition to the neck yoke that is presently worn by Star and

Bright, a smaller, training type yoke was used when they were young. Also, a single yoke may be used to fit the neck of a single ox.

Boardman has acquired or built a variety of machines he uses with his oxen to perform other tasks around his home. These machines are hooked to his team and used when needed. All of these devices are first attached to a forecart which is itself attached to the yoke.

The forecart that Roland Boardman uses with his team he designed and built himself. A full time employee at Maxim Industries in Middleboro, Massachusetts, a company that produces fire apparatus, Boardman is highly skilled with tools and enjoys designing and piecing together the devices he uses with his team. His forecart, which is nothing more than two metal pipes joined diagonally together that ride on a couple of old wheels, was put together in a short period of time. He has also designed and built a wedge-shaped snowplow used to clear the family driveway of snow during the winter, and has assembled a flat pull cart that can be loaded with concrete blocks used to exercise the team. "Building these machines is a hobby," says Boardman. "I don't drink or smoke. I love working around here and enjoy myself, while getting things done at the same time."

Boardman also built the sturdy 12' × 18' barn used to shelter Star and Bright. In it are four stalls, tools, feed and sundry curious relics he has collected over the years. These relics will be used in whole or part to construct other useful gadgets in the future.

The Boardman family maintains a sizable garden which provides them with vegetables for the good part of a year. When the ground needs to be broken, first a plow, then a harrow, are hitched to the forecart. Roland Boardman attaches the forecart to the yoke and leads Star and Bright to the fields for work. He characteristically rests one hand in the pocket of his dungarees and holds the whip with a lash in the other. He walks briskly beside them and covers the area by directing them with his verbal commands. "It's easier to lead them than you might think," he says. "I know of a Connecticut family whose 4-year-old boy drives a pair with a combined weight of over two tons. They're trained to react to the commands and behave better than you'd expect." And when the grounds need to be fertilized Boardman's oxen help take care of this too. They provide about $60 worth of fertilizer each year from their own wastes.

Across the street from Boardman's home is a large, open, grassy field where Star and Bright exercise and eat. Periodically, the oxen and their master will cut the tall grass using a mowing machine that is attached to the forecart. Boardman sees this as another valuable aspect of their versatile nature. "You couldn't run a power mower through that grass." he says. "And the oxen don't need to have their gas tanks filled or their oil changed."

Presently Star weighs 1,600 pounds and Bright weighs 1,400 pounds. On their fourth birthday they should reach full maturity and will then weigh about one ton each. Depending on how hard they work, it costs about $2 per day to keep them which includes the hay and stock primer they eat. Grass from the nearby field and water round out their diet.

"I receive as much enjoyment from owning these two as a boy would a dog," Boardman says, proudly pointing to his oxen. "I intend on always having a pair."

On occasion, Star and Bright are scheduled to appear at a nearby fair or in a parade and will then receive special attention. Each is given a shampoo, a careful and thorough brushing, and if necessary, Boardman will trim their hooves with a hammer and a wood chisel. This trimming is done every few months. Basically this is all the care they require.

Future plans for Star and Bright include entering them in pulling contests which are held during county fairs in many parts of New England. A weighted cart, starting with about a ton, is pulled by the team. Weight is added until the team can no longer pull the cart. The highest number of pounds pulled, which is usually in excess of 20,000 pounds, can draw cash prizes of anywhere from $50 to $150. A list, which includes the dates, times and places of most of these New England fairs, can be obtained by writing to the State Department of Agriculture. The oxen can be easily transported to the fair in a horse trailer.

Boardman has found very little existing literature on the purchase, care and training of a pair of oxen. He believes that talking to fellow oxen owners, and firsthand experience, are the best teachers.

A potential owner should first see anyone who has cows. Actually, cattle outnumber horses by the millions, and obtaining a pair would be relatively easy. There is a large variety from which to choose: Brown Swiss, Durham, Devon, Limousin, Charolais, Chianina and Maine Anjou. For the beginner, some things to look for when selecting an ox are clear eyes, a good frame and girth, a broad back and deep chest and solid hind quarters. A pair of young calves can be bought at whatever the prevailing market price. The market is high now, and calves have been selling from $50.00 to $100.00.

"About the only problem I've ever had with Star and Bright was when they ran into a hornet's nest," Boardman says, rolling his eyes in mock disapproval. "It took me about 30 minutes to slow them down. So far they've never been sick. I have to worm them about once a year, but otherwise, there have been no major difficulties. I imagine it would be pretty tough to get a vet to come out and work on animals of this size, so I'll probably have to handle these things myself when they occur."

Boardman also points out that when driving the oxen care should be taken to prevent getting stepped on. Presumably, 1,600 pound Star could do quite a bit of physical damage if he walked across your feet. James Avery of Massachusetts, however, had much more to worry about when sometime around the turn of the century he owned the largest recorded team in history. His pair had a combined weight of 8,600 pounds.

To move backward in time is not the intention of Roland Boardman. On the contrary, he is quite like many other Americans who are seeking new ways to become more self-sufficient. He has found that not only do Star and Bright give him the satisfying rewards of owning domesticated animals, but they are saving him a substantial number of hard to come by dollars. Unlike the Biblical days when worldly riches were measured largely in terms of herds and flocks, Boardman has discovered unique wealth and enjoyment in just one pair of oxen. "They're good animals," he says. "Some humans are not this reliable."

BEEFALOS

Beefalos are going to pay off for the backyard farmer. People will be able to raise their own beef, cheaply," says Mike Swain who operates "Mainly Beef" with his sister, Mary, in Litchfield, Maine. Mike got started with Beefalos a few years ago by breeding a few Holsteins with Beefalo semen at the large dairy farm where he works. Mary says her ultimate aim is to have fifteen producing purebred Beefalo cows on their 150 acre family farm. Mike and Mary want to produce inexpensive, quality beef and they believe Beefalos are the perfect breed for this. If they're right, they may do very well in the process, since they sold their first half-blood Beefalo cow with her three-quarter blood heifer calf for $4,000 at a Beefalo production sale.

So what's all the hullabaloo about Beefalos? Is the promotional literature that's been flying around really true? Since the story came out a few years ago, I've always been of the opinion that it was too good to be true, but a recent visit with Mike and Mary Swain has at least changed me into a passive optimist about the future of this new breed, and with it the future of beef, forage-fed beef.

It all started about fifteen years ago when Bud Basolo, a California cattle breeder, was finally successful at producing a fertile hybrid between domestic cattle and American bison. Cattlemen had been trying this for years, but the hybrids were always infertile, as are mules (donkey-horse hybrids). Basolo's first breeding animal was a holstein-bison cross. I'm told it looked rather weird, but it was fertile. That's all that mattered. From this original animal Basolo kept cross breeding and cross breeding until he came up with a combination of bison and beef cattle breeds that would breed true and be of uniform, beef cattle-like conformation. He finally found it, and the Beefalo breed is three-eighths American bison and five-eighths beef breeds. Almost all the bulls in service are three-eighths charolais, one-quarter hereford and three-eighths bison.

The primary characteristic of this breed which has created such a stir is that it needs no grain. It thrives on roughage, such as grass, hay and silage, alone. The calves will reach slaughter weight of

Fig. 6-1. Illustration of the parentage of a Beefalo.

1,000 pounds at twelve to fourteen months of age, with absolutely no grain. It takes an intensive grain feeding program to get most cattle breeds to 1,000 pounds by fourteen months. Left on low-quality forage alone, most cattle would be at least two years old by the time they reached slaughter weight.

Mary says they won't even eat grain if they have the choice of grain or hay. "I've demonstrated this to people at fairs. I gave them hay and a handful of grain. They wouldn't eat the grain." And these were only half-breed Beefalo she was talking about. Most of the characteristics of the breed are carried by animals that are one-half Beefalo and one-half other cattle breeds, in other words, only three-sixteenths bison.

Mary has raised a half Angus, half Holstein calf together with a half Beefalo calf to compare some of the costs. She says the Angus-Holstein calf consumed $350 worth of grain over thirteen months. The Beefalo consumed a ton of beet pulp ($120) in the same period, but no grain. They both ate hay and the Beefalo undoubtedly ate more, but hay is a lot cheaper than grain.

Bud Basolo has been quoted as saying he didn't want to be known as a man who produced a new novelty breed but as the man who made "cheap beef" possible. It is estimated that Beefalo cuts will be 25 to 40% cheaper than the average choice beef. The Canadians must believe this because the only bull that Bascolo has sold so far went to a Canadian sydicate for 2.5 million dollars.

"Beefalos are tremendous foragers. When the other cattle are in the pasture, the Beefalos often forage in the woods," says Mike Swain. That's one of the problems with modern cattle breeds. They've been bred for easy management in confinement and are often not good foragers on their own. One of the best studies which I've seen which explains why beefalos utilize low quality forage better than normal breeds of cattle came from Poland. Researchers compared the digestibility of nutrients in feed by European bison (closely related to American bison) and cattle. Bison digested protein, fat and lignin more efficiently than cattle. Cattle equal bison in cellulose digestion and the digestion of readily soluble carbohydrates. Lignin is a very indigestible component of low quality forages. The bisons' ability to digest lignin, protein and fat more efficiently than cattle explains why they can utilize forage so well.

As the cost of grain, energy and intensive management continues to increase, meat animals that can live on low quality foods and are good foragers will be looked at more favorably. Maybe the great American feedlot at will go down in history as an archaic, inefficient method that provided rich folks with beefsteak in the twentieth century.

Another characteristic of the breed that makes them so rugged is their wooly bison undercoat. Again, even the half Beefalos have this characteristic. Beefalos have about three times as many hairs per square inch of hide as regular cattle breeds. "When you run your hand up the coat of a regular cow and part the hairs, you can see skin between the hairs. I can always tell a Beefalo because you can't see any skin," says Mike. Beefalos also have longer straight guard hairs which extend out beyond the wooly undercoat. This heavy winter coat makes Beefalos very tolerant to cold, windy weather. "Beefalos can stand the cold or heat. They slick right down in the summer and have a very short coat."

Besides the benefits which the animals receive from this wooly coat, the owners are receiving a considerable bonus too. Prime hides, from animals slaughtered in late fall and early winter, have sold for up to $300 each. They've been made into buffalo robes, or rather, Beefalo robes.

Beefalos are gentle too, or as gentle as any other cattle. Bison are known for being good fence breakers. It's been attributed to their migratory habits. When a herd of them decides it's going to travel, they travel, paying little attention to fences. Bison bulls are known to be a bit nasty, too, but that's nothing new. Dairy breed bulls are undoubtedly the most dangerous animals in North America, in terms of the number of people they've killed. Mike and Mary say there are no special handling problems with Beefalos. They are as gentle as their Herefords and Holsteins.

If this is the beef animal of the future, it has to be good to eat and produce the cuts and types of meat that people are used to eating. According to reports, the meat is excellent. It is high in protein, 19-22%, as compared to most beef which is only 10-12%. It also averages only 7% fat, lower than most beef. Half-blood Beefalos dress out at 62%, and they have more lean meat than regular cattle. The Texas A&M University grading service slaughtered a large sample of animals, and their results confirmed the promotional literature. Animals (half Beefalo) came from all parts of the country and averaged twelve months of age and 955 lbs. They dressed out to 62% and produced more lean cuts and more lean ground beef than regular cattle. The meat appears no different than regular beef with the exception that there is less exterior fat and less marbling with fat in the meat. Several cattlemen who raise

Beefalos have been noted as saying that within ten years all slaughter cattle will have some bison blood. Scientists who evaluated the grading study from Texas A&M tended to agree with this.

Even with all these impressive statistics, most livestock experts are still taking a wait-and-see attitude. As far as I know, however, none of them have any facts which refute the promotional literature put out by the Beefalo associations.

There are two Beefalo associations—one in the west and one in the east. Beefalo East has headquarters in Virginia and New York State, and registration in the east is by the American Beefalo Association, Louisville, Kentucky. If you are interested in breeding cattle, you must obtain semen from franchised dealers. Solo Beefalo, Inc., Ridge Road, Fairfield, ME 04937 has the franchise for the six New England states. The price is $12 per ampule for up to nineteen ampules, with lower prices for larger orders. There are about thirty beefalo bulls in service now from which to select semen. There are three different families available and with careful family selection one can avoid inbreeding.

As for breeding difficulties, Mike Swain says there are none. Mike confirmed what the literature said; the conception rate from beefalo semen is comparable to other breeds. Mike is an experienced artificial inseminator, since he inseminates Holsteins at the large dairy farm where he works. Another good thing about beefalos is that their calves are small, only forty to sixty-five pounds, so cows have little calving difficulty. The promotional literature says "over 10,000 births—not one assisted." I can't quite believe that but at least there are no notable difficulties. Cattlemen are especially interested in ease of calving since it has been a problem in crossbreeding programs between European and English beef breeds. Certain crosses produce large calves and as many as ten or twenty percent of the calvings must be assisted.

It would be difficult for a person with one or only a few cows to breed his cows to Beefalo at present. The semen must be obtained and stored in a liquid nitrogen tank until needed. Most artificial insemination is done by technicians of the Eastern Artificial Insemination Association. Eastern AI will not allow its technicians to handle semen from bulls which are not within the association, with semen handled to their specifications. They have no Beefalo bulls at present. Mike's suggestion for the small farmer is to make a deal with the nearest large dairy farm which has a liquid nitrogen tank to store the semen. Then make arrangements with a vet or a local A.I. technician to do the breeding at the proper time.

Hopefully this procedure will be simplified in the future because, as Mike pointed out, there might be a real future for the small farmer in raising Beefalos. Especially in New England, where hay is still abundant, but which is a long way from the nearest grain-growing region.

Right now Mike and Mary are breeding a small herd of hereford cows with Beefalo semen. They also have a half-Holstein/half-Beefalo heifer to breed. On 50 tillable acres they raise 35 acres of hay and 15 acres of corn silage. They make a cooperative deal with Alice Wheeler, Mike's boss, and allow her to put in the 15 acres of no-till corn silage, of which Mike and Mary get to keep a portion to feed their cattle. In return they are allowed to use Alice's equipment to make their hay. Mike feels that small farmers should look into this type of cooperative agreement with nearly large farmers who own equipment. Mary is obviously proud of the land improvements they've made by land leveling, drainage and reseeding under ASCS cost-share programs. With the dreamed-of herd of fifteen purebred Beefalos they'll have an impressive spread, and be able to provide people with inexpensive, forage-fed beef.

HIGHLAND CATTLE

We have had a few cows around Amen Farm ever since we settled here in 1959. In the beginning, having big ideas about dairying, we bought a Guernsey cow. She was what is known, colloquially, as a three teater, which meant that only three quarters of her udder was operable, the remaining quarter having been blanked out by mastitis. She was a pedigreed animal and we bought her out of a dairy herd. The only reason she was for sale was because of her disability. However, she provided all the milk, and butter, and cream and cottage cheese we could use, sell, and give away. Her cream was so thick that a fork would stand up in it and I blame her for my high cholesterol count.

We had Cindy for a long time and she presented us with a succession of beautiful calves. Her sons we slaughtered at six weeks for veal and her daughters we grew on for additional milkers. Cindy was a gentle, friendly animal. After some years, as the young heifers matured and freshened, we sold Cindy in favor of her children. So far as I know she never bred again but continued to give a decreasing amount of milk for a couple of years. I am not particularly soft-hearted but I am glad I did not have to make the decision to dispose of her as "canner grade" beef.

Fig. 6-2. A Highland Cattle sketch.

Being entranced by the beautiful colored posters published by the Swiss Tourist Bureau, picturing cattle up to their udders in buttercups and daisies in Alpine meadows, we added a Brown Swiss to our herd. She was enormous and gentle, and not brown at all, but exactly the color of a field mouse. She was the easiest milker I have ever known. All you had to do was to slap her udder and she would begin to let-down. We kept her only a short while because after being accustomed to golden Guernsey milk hers looked and tasted like chalk and water.

Finding we had far more dairy products than we could use or dispose of we cut back to one milch cow and added a Black Angus calf to our establishment. Actually, I bought her on impulse. We were at a cattle auction at Exeter when a little Angus heifer was brought into the ring. When the handler poked her with his cattle prod she turned and chased him out of the enclosure. She looked small and defenseless and I admired her spunk so I bought her for $46.00. We named her Bonnie and had her for five or six years. After she got going she added a calf every year on schedule. The only trouble with her was that she was as hard to handle at home as she had been in the ring. I thought I had her tamed but one evening as I was bedding her down she took aim and kicked me clear across the barn.

In those days we did not keep a bull. There was a gentlemen in Bucksport who acted as proxy. On a couple of hours notice he would hurry over with his little black bag and give you the choice of half a dozen pedigreed bulls of different breeds at $2.00 a go. If the first one did not take he would give a second one free. He did not often miss. Alas, that service is gone, as there are no longer enough cows in the area to make it worthwhile.

Currently we have three Scotch Highland cattle, a bull for the reason stated in the preceeding paragraph, a bred heifer, and a steer. The heifer is about to drop her calf momentarily, which is why we put her in the pasture this morning. The bull and the steer are there too, playing in the cold April sunshine. My experience with the breed is limited, as I have had them for only about six months, but so far they have been a delight. They are gentle, easy to handle, make a living on poor forage, and need no protection in our tough Main winters other than a shelter to keep out of the bitterly cold blizzards. In ordinary winter temperatures, above zero, they merely lie in the snow and chew their cuds.

My decision to try Highlanders arose out of an effort to get some small, easily handled beef cattle. They were really a second choice because what I wanted were Dexters that are even smaller than Highlanders. Dexters are an Irish breed that are raised as family cows, as indeed were the Highlanders or, as they are sometimes called, Byloes. On reflection I believe I was lucky as the Byloes can stand our severe climate better than the Dexters would have been able to do.

Scotch Highland cattle with their long shaggy coats and spreading horns look exotic and we have a lot of visitors to see them. We also have many inquiries about them from young 20th century homesteaders. Almost always the first question from the latter is "Can you milk them?" Well, any cow can be milked, but the question is whether the result is worth the trouble. Most beef animals give rich milk but not being bred for that purpose they do not do so in quantity and most of them look with disfavor upon being milked by other than their calf. There are dual purpose breeds such as the Milking Shorthorns but, generally, raising cattle in the United States is a highly sophisticated business and almost all dairy herds are composed of Jerseys, Guernseys, Holsteins, Brown Swiss, etc. while the commonest beef breeds are Angus, Herefords, Santa Gertrudas, etc. While dairy breed calves (bulls) are slaughtered for veal, and old cows for canner-grade meat, all high grade beef comes from animals bred solely for that purpose. Anyone familiar with cattle can distinguish dairy and beef breeds merely by their bodily configuration.

I have also been asked if a Highlander could be used as a draught animal, as an ox. Again there would seem to be no reason why not, certainly they are biddable and could soon be taught to work in harness, but they are small and could not pull as would a big Holstein.

I doubt that raising Highlanders would be a profitable commercial venture in this part of the country, but for my purpose they are an excellent source of home grown quality beef, and the emphasis is upon quality. They are a good breed for the amateur because of their ease of handling, their ability to get along on rough forage, their good mothering instinct, and the fact that they are not subject, as are many milch cows, to milk fever.

There is a lot more that could be said on this subject but I would like to emphasize two points. Do not attempt to keep a cow, any sort of cow, unless you are prepared to stay at home to care for it. The other is (unless you are rich) not to get involved unless you can raise your own hay or get it free from someone else for the making and bailing. Even Highlanders cannot live on cedar tips like deer, and hay last winter was $1.50 a bale, more near the big cities, and at that price you can buy prime beef cheaper than you can raise it.

Visitors are always welcome at Amen Farm and if I am around, which I usually am, I shall be glad to dispense what wisdom I have.

Chapter 7

Equines

Relatively few horses are inspected and evaluated by experienced judges. Most of them are evaluated by persons who lack experience in judging but who have a practical need for the animal and take pride in selecting and owning a good sound horse.

HOW TO BUY A HORSE

As a guide I would like to describe the four basic methods you can use in buying a horse. You should use more than one method whenever possible in choosing your horse.

Pedigree: Selecting animals by pedigree, or on the basis of their ancestors, is of special importance where animals are either too thin or so young that their individual merit cannot be determined accurately. Pedigree may be the determining factor when selection is made between animals of comparable individual merit.

Show-Ring Winnings: Because training plays an important part in the performance and show-ring winnings of light horses, this method of selection is of less value from a breeding standpoint than with other classes of farm animals. However, performance of a horse in the show ring can be a valuable criterion in indicating its utility.

Performance and Progeny Testing: Performance testing refers to testing or evaluating animals by measuring their actual performance—for example, by timing their speed over a certain distance. Progeny testing refers to the practice of selecting animals on the basis of the merit of their progeny.

Type of Individual: Selecting by individual excellence of body conformation and performance of the animal is the best single method of obtaining suitable horses. When animals are selected for breeding purposes, however, certain additional facts should be taken into consideration.

To select a sound horse by type of individual—the method of selection used by the majority of people—you should:

• Know the names of the various anatomical parts. Master the language that locates and describes the parts of a horse. Know which of these parts are of major importance and what comparative evaluation to give the different parts.

• Know what you want. Have an ideal in mind. Be able to recognize desirable characteristics and common faults.

• Follow a definite procedure in examining. Size up a horse by following a logical procedure, such as indicated in the table; look the horse over in front view, in rear view, in side view, and in action. Check for soundness. In this way you will not overlook anything and you will find it easier to retain observations as you make them. When you are looking over several animals at the same time, keep them at a distance to secure a panoramic view.

• Make a sound evaluation. Evaluate the animal on each point listed under "what to look for," and keep common faults and your ideal type in mind. If several animals are involved, rank them in your mind by their rating on important points.

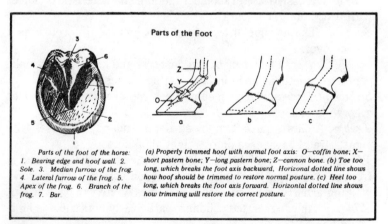

Parts of the Foot

Parts of the foot of the horse: 1. Bearing edge and hoof wall. 2. Sole. 3. Median furrow of the frog. 4. Lateral furrow of the frog. 5. Apex of the frog. 6. Branch of the frog. 7. Bar.

(a) Properly trimmed hoof with normal foot axis: O—coffin bone; X—short pastern bone; Y—long pastern bone; Z—cannon bone. (b) Toe too long, which breaks the foot axis backward. Horizontal dotted line shows how hoof should be trimmed to restore normal posture. (c) Heel too long, which breaks the foot axis forward. Horizontal dotted line shows how trimming will restore the correct posture.

Fig. 7-1. Hoof anatomy and how to tell if it is trimmed properly.

Common Defects In Way Of Going

The feet of a horse should move straight ahead parallel to an imaginary center line drawn in the direction of travel. Any deviation from this way of going constitutes a defect. Some other defects are:

Cross-firing: A "scuffing" on the inside of the diagonal forefeet and hindfeet, generally confined to pacers.

Dwelling: A noticeable pause in the flight of the foot, as though the stride were completed before the foot reaches the ground; most noticeable in trick-trained horses.

Forging: Striking forefoot with toe of hindfoot.

Interfering: Striking fetlock or cannon with the opposite foot; most often done by base-narrow, toe-wide, or splay-footed horses.

Lameness: A defect detected when the animal favors the affected foot when standing. The load on the ailing foot in action is eased and a characteristic bobbing of the head occurs as the affected foot strikes the ground.

Paddling: Throwing the front feet outward as they are picked up; most common in toe-narrow or pigeon-toed horses.

Pointing: Perceptible extension of the stride with little flexion; likely to occur in the long-strided thoroughbred and standardbred breeds, animals bred and trained for great speed.

Pounding: Heavy contact with the ground instead of desired light, springy movement.

Rolling: Excessive lateral shoulder motion; characteristic of horses with protruding shoulders.

Scalping: The hairline at top of hindfoot hits toe of forefoot as it breaks over.

Speedy Cutting: The inside of diagonal fore and kind pastern make contact; sometimes seen in fast-trotting horses.

Stringhalt: Excessive flexing of hind legs; most easily detected when a horse is backed.

Trappy: A short, quick, choppy stride; a tendency of horses with short, straight pasterns and straight shoulders.

Winding or Rope-walking: A twisting of the striding leg around in front of supporting leg, which results in contact like that of a rope-walking artist; this often occurs in horses with very wide fronts.

Winging: An exaggerated paddling, particularly noticeable in high-going horses.

How To Measure The Horse

Normal pertinent measurements are height, weight, girth, and bone.

Height: The height of a horse is the vertical distance from the highest point of its withers to the ground when the animal is standing squarely on a level area. The unit of measurement used in expressing height is the "hand," which is four inches. A horse measuring 62 inches is said to be 15-2 hands high (15 hands and 2 inches).

You can estimate a horse's height if you know the exact number of inches from the level of your eyes to the ground. Knowing this, all you need do is stand beside the animal's front limbs and look at the highest point of the withers; in this way, you can estimate the horse's height rather closely.

Weight: Although there are ways of estimating weight, it is best to use scales.

Girth: Girth is a measure of the circumference of the chest behind the withers and in front of the back.

Bone: Size of bone usually is determined by placing a tape measure around the cannon bone halfway between the knee and fetlock joints. This measurement is in inches.

How To Determine Age

The lifespan of horses averages 20 to 25 years, about one-third that of man. Horses generally are at their best between three and twelve years of age. This may vary because of individual differences in animals or because of differences in the kind of work they do. The age of horses is, therefore, important to breeder, seller, and buyer.

The approximate age of a horse can be determined by noting time of appearance, shape, and degree of wear of temporary and permanent teeth. Temporary, or milk, teeth are easily distinguishable from permanent ones because they are smaller and whiter. The best way to learn to determine age in horses is by examining teeth of individual horses of known ages.

A mature male horse has 40 teeth. A mature female has 36. Quite commonly, a small, pointed tooth, known as a "wolf tooth," may appear in front of each first molar tooth in the upper jaw, thus increasing the total number of teeth to 42 in the male and 38 in the female. Less frequently, two more "wolf teeth" in the lower

Teeth of a Horse

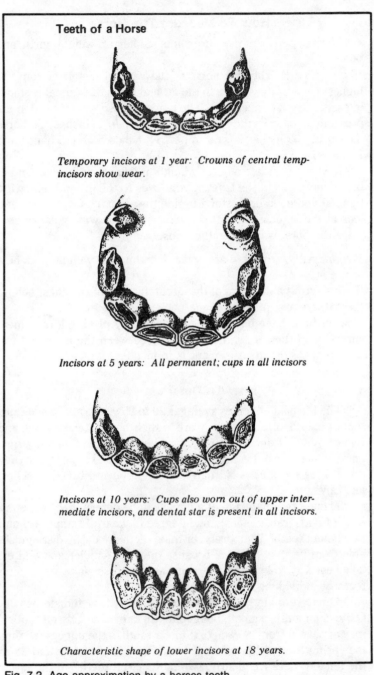

Temporary incisors at 1 year: Crowns of central temp-incisors show wear.

Incisors at 5 years: All permanent; cups in all incisors

Incisors at 10 years: Cups also worn out of upper intermediate incisors, and dental star is present in all incisors.

Characteristic shape of lower incisors at 18 years.

Fig. 7-2. Age approximation by a horses teeth.

jaw increase the total number of teeth in the male and female to 44 and 40, respectively. A foal of either sex has 24. Even experienced horsemen cannot determine the age of an animal accurately after it is 12 years old. After this age, the teeth change from oval to triangular and they project or slant forward more and more as the horse becomes older.

The animal's environment can affect wear on the teeth materially. Teeth of horses raised in a dry sandy area, for example, will show more than normal wear; a five-year-old western horse may have teeth that would be normal in a six-to-eight year-old horse raised elsewhere. The teeth of cribbers also show more than normal wear. It is hard to determine the age of such animals. The age of a horse with a parrot mouth, or undershot jaw, also is difficult to estimate.

Blemishes And Unsoundnesses

An integral part of selecting a horse lies in your ability to recognize common blemishes and unsoundnesses and your ability to rate the importance of each. A thorough knowledge of normal, sound structure makes it easy to recognize imperfections. Any abnormal deviation in structure or function of a horse constitutes an unsoundness. From a practical standpoint, however, a differentiation is made between abnormalities that do and those that do not affect serviceability. Blemishes include abnormalities that do not affect serviceability—such as wire cuts, rope burns, nail punctures, shoe

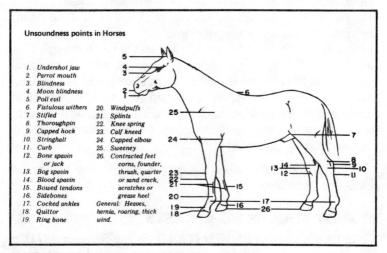

Fig. 7-3. Points to observe to determine soundness.

boils, or capped hocks. Unsoundnesses include more serious abnormalities that affect serviceability. Consider the use to which you intend to put the animal before you buy a blemished or unsound horse.

Stable Vices

Stable vices are bad habits of the horse in confinement. They may detract from the value of the animal.

Cribbing: A horse that bites or sets his teeth against the manger or some other object while sucking air is said to be cribbing. This causes hard keeping and a bloated appearance. Horses with this vice are subject to colic. A common remedy for cribbing is a strap buckled snugly around the horse's neck in a way that will compress the larynx when the head is flexed, but that will not cause any discomfort when the horse is not indulging in the vice.

Halter Pulling: This term is applied to a tied horse that pulls back on its halter rope.

Kicking: A true stable kicker apparently kicks just for the satisfaction it gets out of striking something with its hind feet. Unusual excitement or injury occasionally causes a gentle horse to kick.

Tail Rubbing: Persistent rubbing of the tail against the side of the stall or some other object is objectionable. Parasites, such as lice or rectal worms, may cause this. A "tail board" or parasite control helps break animals of this habit. A tail board is a board projecting from the wall of the stall high enough to strike just below the point of the buttock, instead of the tail, of the rubbing horse.

Weaving: A horse's rhythmic swaying back and forth while standing in the stall is known as weaving.

Bolting: Horses that eat too rapidly are said to be "bolting." This can be controlled by adding chopped hay to the animal's grain ration, or by putting stones at least as big as baseballs in its feed box.

Other vices: Other vices often difficult to cope with, especially in older animals, are: balking, backing, rearing, shying, striking with the front feet, a tendency to run away, and objecting to harnessing, saddling, or grooming.

Most of these bad habits are caused by boredom and by incompetent handling.

MY FIRST TEAM OF WORKHORSES

In the first issue of *FARMSTEAD* (Spring-Summer 1974), a greenhorn (myself) tried a single workhorse named Lady for farm

What To Look For	Ideal Type	Common Faults
Front view:		
1. Head	1. Head well proportioned to rest of body, refined, clean cut, with chiseled appearance; broad, full forehead with great width between eyes; jaw broad and strongly muscled; ears medium size, well carried, and attractive.	1. Plain headed; weak jaw.
2. Femininity or masculinity	2. Refinement and feminity in brood mare; boldness and masculinity in stallion.	2. Mares lacking femininity; stallion lacking masculinity.
3. Chest capacity	3. Deep, wide chest.	3. Narrow chest.
4. Set of front legs	4. Straight, true, and squarely set front legs.	4. Crooked front legs.
Rear view:		
1. Width of croup and through rear quarters	1. Wide and muscular over croup and through rear quarters.	1. Lacking width and length over croup and muscling through rear quarters.
2. Set of hind legs	2. Straight, true, and squarely set hind legs.	2. Crooked hind legs.
Side view:		
1. Style and beauty	1. High carriage of head, active ears, alert disposition, and beauty of conformation.	1. Lacking style and beauty.
2. Balance and symmetry	2. All parts well developed and nicely blended together.	2. Lacking in balance and symmetry.
3. Neck	3. Fairly long neck, carried high; clean cut about the throat latch; with head well set on.	3. Short, thick neck; ewe necked.
4. Shoulders	4. Sloping shoulders (about a 45° angle).	4. Straight in shoulders.
5. Topline	5. Short, strong back and loin, with long, nicely turned and heavily muscled croup, and high, well-set tail; withers clearly defined and of same height as high point over croup.	5. Sway backed; steep croup.
6. Coupling	6. A short coupling as denoted by last rib being close to hip.	6. Long in coupling.
7. Middle	7. Ample middle due to long, well-sprung ribs.	7. Lacking middle.
8. Rear flank	8. Well let down in rear flank.	8. High cut rear flank or "wasp waisted."
9. Arm, forearm, and gaskin	9. Well muscled arm, forearm, and gaskin.	9. Light-muscled arm, forearm, and gaskin.
10. Legs, feet, and pasterns	10. Straight, true, and squarely set legs; pasterns sloping about 45°; hoofs large, dense, and wide at heels.	10. Crooked legs; straight pasterns; hoofs small, shelly, and contracted at heels.
11. Quality	11. Plenty of quality, as denoted by clean, flat bone, well-defined joints and tendons, refined head and ears, and fine skin and hair.	11. Lacking quality.
12. Breed type (size, color, shape of body and head, etc.	12. Showing plenty of breed type.	12. Lacking breed type.

Fig. 7-4. Parts to check and a comparison of good and bad.

and woods work. Since then, another horse, a 1600 pound Belgian mare named Trixie, has been paired with our old roan mare to make a team. If Lady had given such good service, why did we go to the added expense and trouble of acquiring an additional horse, and the different harness and equipment required for a team? What benefits might compensate for the added trouble and expense? What is it like to work a team as compared with a single horse?

It was probably natural that, having acquired some experience

153

with one horse, I should want to see what it was like to use two. Perhaps it is more exciting to see a good team pull together, and there is the added challenge to the teamster of making certain that the horses pull smoothly rather than seesaw. Perhaps, too, the fact that Trixie had spent part of the previous winter in the barn with Lady, that they had been worked in company with each other, and finally that in the spring Trixie became available—all had something to do with the decision to use a team. As a matter of fact, Trixie's owner and I had often speculated as to what kind of a team they would make, and one day, we had even reined them in side by side and had been impressed with how well matched they seemed in size and gait.

It might seem that these were my only reasons for taking such a step, but this was far from being the case. All along, there were good solid reasons for going to a team operation. A neighbor, with a team which he uses on his woodlot, had told me that only with a team would I have the necessary power to get the wood to a point where it could be loaded on to a truck. From almost every direction on our place, wood must be taken uphill before it can be loaded. In addition, slopes have a way of icing, and two horses as a team learn to lean a little on each other where a single horse will slip and fall.

There were other reasons why a team would meet our needs better than a single horse, I thought. In field work, there is some machinery which comes fitted only for team use, such as a manure spreader, grain drill, two-row planter, and riding or sulky plow. Thus a team would enable me to perform a greater variety of tasks, and some double machines such as a mowing machine with a six-foot cutter bar are far more efficient than their single horse counterparts such as a single horse mower with a three-foot bar. Some tasks are simply beyond one horse, such as pulling a hay wagon with a wheel driven hayloader behind. Two horses are company for each other and therefore are usually more contented and easier to deal with. A team is less likely to take off, because with double reins connecting them, they both must go at the same time and in the same direction or not at all, (though I've had it happen, both single and double).

My reasons for trying a horse, originally, were that a horse would be more compatible with the goals of our farm than a tractor. However, it seemed worthwhile also to see whether a greenhorn could learn to use a team, for any significant revival in the use of work horses as a low energy alternative would have to involve the use of teams on larger farms.

The day came when we were ready to try Trixie and Lady together. Harness parts had been secured with which their single harnesses could be converted to double horse use. In addition, and more important, I had an experienced horseman, Trixie's former owner, to help me out. Lady, at 17, was still in good shape, though Trixie outweighed her by a couple of hundred pounds and was half her age. On the other hand there was not so much difference in size and conformation as to make it appear difficult for them to work together. (Sometimes the most ill-assorted seeming pair will work out very well as a team.) Differences in strength and pulling power may be compensated by adjustment of the trace chain length and by moving the point at which the evener is pinned toward the stronger horse. In any case, we harnessed them, hitched them to a stone boat, and held our breath. Somehow the chemistry was right, and the two walked off as though they had worked together for years. By that afternoon, they were hard at work pulling the manure spreader. Better yet, very soon we were able to use them on the sulky (two way) plow, something which often must be postponed for a new team because of the extra precision required. One factor which must have helped to make this transition as smooth as it was was that each horse had had team experience at some time in the past.

The team got off to a good start, harrowing and spreading a six acre field which had been plowed earlier that spring before the team was available for such tasks. I had little to complain about, or so it seemed, as the team finished off work on the field by pulling the grain drill which seeded oats and timothy at the same time. The horses were obedient, easy to handle, and were good about standing when left alone. So I was looking forward to haying season in which the new team could play a major part, but it was not to be. The difficulties which we experienced early that summer, show that owning and working horses is not always an easy thing. Some things are learned only through rather painful experience.

The incidents which limited our haying with the horses were two, the first of which put Trixie out of commission for several weeks. The second almost cost us Lady's life, and it resulted from my own carelessness. Early in haying season, after having had a chance to mow only a little with the team, Trixie turned up lame. To consult the section on lameness in a horse book is to open Pandora's box. There are pages and pages devoted to the subject, and none of them did us any good so far as a diagnosis was concerned, until by tapping the hoof lightly with a hammer we determined that Trixie had a hoof infection. The cause was either a horseshoe nail in the quick or a small piece of gravel which had worked its way up between the inner and outer hoof layers. In any case, there was little to do but remove her shoe, give her injections to combat pain and infection, and soak the hoof daily.

Meanwhile, haying continued with one horse. We pulled the mower with a tractor, but used Lady to pull the hay rake and the wagon, sometimes alternating her from one to the other. Then one night, I failed to latch the stall door properly, and Lady was lost to us for a week. The two horses got into the main part of the barn and into various kinds of grain that were stored there. If you ever get horses of any kind, for goodness sake make certain that they never get into grain. It is a rare horse who knows any restraint where grain is concerned, and stomach overfilled with the heating grain, the horse cannot regurgitate what he has eaten. If a blockage develops and the grain cannot pass through, the horse will die. In addition, the horse may develop black water, a muscular involvement which causes lameness which may be permanent. Beyond that, a horse's hooves may founder or delaminate up to several months after such an incident.

What followed was a nightmarish experience which included getting Lady up (it took four of us), walking her for several hours to get things moving, we hoped, and then staying up with her all night to make certain she did not lie down for very long. My penance was capped by a trip to Belfast the next morning to get medicine. By this time it was clear that Lady was going to survive, but right along she showed clear signs of blackwater in a rear leg. So, when late the next afternoon Lady got up and walked off without showing any signs that anything had happened, we cheered our good luck at getting off so easily.

We finished haying with one horse, and by early August, Trixie had recovered from the hoof infection and had been shod again. And here, by the way, might be the place to discuss hoof quality. Generally, the horn of horses hooves is either dark or light, and the color is usually an indication of how hard the hoof material is, and of how well it will hold shoe nails. Lady's hooves are dark and they are so hard that her shoes usually stay on the regular three month term with very little re-nailing required. (This is the kind of hoof to look for when you are buying a work horse. And make sure that no one has used some artificial coloring to obtain this effect, either.) Trixie's hooves were quite light, and the horn does not hold nails well. This also means that she must be kept shod all the time, for otherwise she will break up her hooves, making it more difficult to nail a shoe on.

It was for this very reason that it had been difficult to reshoe Trixie after the hoof infection, and the farrier told me that we would be lucky to keep the shoes on for six weeks. Sure enough, the first shoe came off, and I called to see about having it put back on. Now, it is good to know how to do this sort of thing yourself, so when it was suggested that I give it a try, in fear and trembling I did so. That first time it took fully three hours to reset that shoe, and I was in a lather by the time it was done. What took so long and was so exhausting is now more or less routine.

Although I had now learned to keep Trixie's hoof problems under control, I had not reckoned with the consequences of the hoof infection which were to ensue almost a year later. Where the infection had broken out above the hairline there later developed a horizontal crack which appeared on the side of the hoof as it grew down. This was perfectly normal, but represented a shoeing problem, as we had to make certain that, in shoeing, the crack would be prevented from continuing around the hoof with the possibility of breaking away much of the outer horn. We managed to work around this problem until late in the spring when the crack had grown down almost all the way. Then what we had feared happened, the shoe loosened, and before we could take steps to reset it, it broke away, taking much of the outer hoof material with it and leaving very little to nail a shoe onto. There was very little protection left for the inner sensitive portion of the hoof, so it was either come up with a special shoe that would protect the hoof while it grew out, or face the distinct possibility that she would go lame and be unable to work.

I won't go into all the details of making a model shoe out of wood and sheet lead, fitting it to her foot, and then duplicating this arrangement as nearly as possible by having clips welded to the shoe. To these clips were attached sheet metal clips which practically covered the hoof, and which bolted together across the front. This special shoe with a few changes and reset a couple of times has been on ever since—with the result that the hoof is now practically normal. Part of the general hoof problem here is related to diet, and we are feeding a diet additive called "Drive" to promote hoof growth. Thus Trixie's hooves are growing out, and I do not see this as a serious problem from here on. I tell this story not because I think that many people will encounter this sort of thing, but because it is a good idea to check hoof color and condition when you look at a work horse.

After fourteen months or so of uninterrupted service by the team, it is time to review what they have accomplished. Plowing and harrowing, especially the latter, are the most difficult tasks they face, and certainly they wouldn't win any races with a tractor. In 1974, we plowed four acres of new ground with horsepower and seven acres with a hired tractor, but of the latter amount, five were plowed in the early spring before the team was available for service. This year the team has been used to plow six acres of new ground, some away from our place. A borrowed tractor with tiller was used to break up five acres of blueberry ground which, in view of the difficulty of the task and the time limit of a government cost shares program, could not have been done by the horses. In addition, the horses are now hard at work turning over land which was under cultivation this year, having done some three acres so far. This is a task which horses accomplish more easily and rapidly than breaking new ground. We feel that the sulky plow we use compares favorably with a tractor plow in turning a neat furrow. We stick pretty much to horse harrowing, too, because the difficulty of the task is offset by the fact that with about five feet covered with each pass, it doesn't take long to do a field. Moreover, we avoid that compaction of the sub-soil which results from tractor use. Also, there are times when we can get on our fields with horses to plow and harrow when it would be too soft and wet for a tractor. Now that all the tillable ground on the farm, except horse and sheep pasture, has been broken at least once, we foresee using horses exclusively (except for outside fields, where for logistical reasons, we will probably have to rely on a tractor some of the time).

While the ground is being prepared for planting by use of the plow, disc harrow and spring tooth harrow, the horses are also used to pull the manure spreader, perhaps the single most important piece of equipment on the place. We try, when bringing back an old field, to fertilize all at one time by putting an even layer each of rock phosphate and granite dust on top of each spreader load of hen manure, so that the contents of the spreader are, we hope, mixed and distributed evenly over the field.

With the ground all prepared, it is time to plant the crops. This year we used a horsedrawn grain drill to plant eight acres of oats. It is a pleasant feeling to sit on the back of the drill as the horses move steadily along. All the hard work of plowing and harrowing is over, and this is easy and relatively fast. The drill covers over seven feet at a pass. You can see the grain spilling out of the metal shutes into the furrows just made by the discs only to disappear instantly under the heavy plank drag which buries the seed. It takes about thirty minutes to drill an acre of grain, and with the newer grain drill it will be possible to fertilize as you sow if desired. Another occasion this year on which we used the drill was to sow a newly tilled piece of blueberry ground with winter rye as a cover crop for green manure which we hope to plow under next spring.

Last spring the horses found themselves planting a little over two acres of dry beans with a two row "King of the Cornfield" planter. As the name implies, this machine will also plant corn as well as other row crops. Here the rate is about forty minutes to the acre, and the machine is easily managed by the team. As with the newer drill, fertilizer can be put down with the seed. The planting done, the horses pulled a riding cultivator through the bean patches to keep the weeds down. The horses and the two banks of cultivators straddle the row, and in addition to guiding the horses with the reins, the teamster must steer the cultivator with foot pedals. Occasionally, if hand and foot coordination are not good, you are likely to wipe out part of a row of corn of beans before getting back on course again. (It helps if the teamster guided the planter on a straight course to begin with.) Our plans call for expanding the acreage in dry beans, for we feel that the horses will make it relatively easy to plant and take care of a much larger crop than before.

With a sound team for haying this year we got a better idea of what horses could do. Using a double horse mower we cut eight acres of hay on the farm twice. Each cutting took less than a day. In addition, the horses did almost all of the tedding and raking. For picking up the hay in the field we used a hay loader for the first time this year. (With this marvelous machine you can even pick up the hay without windrowing it if you are pressed for time.) We established that the horses could pull the hay wagon with loader on behind so that all that was needed was someone to build the load on the wagon as the hay came cascading over the back. For unloading hay in the barn we set up a hay fork on block and tackle. This too may be horse powered, although last summer we used a pickup truck. Thus the entire haying operation may be performed by horses. In practice, however, we found that in extremely hot weather and if we were pressed for time to get the hay under cover, it was better to use the hayloader behind the truck. I might add that putting up hay with a hayloader and hayfork compares favorably with the modern method of baling so far as time and effort are concerned, unless in addition to the baler you own equipment to handle the bales automatically both in the field and in the barn. We plan to use horsepower, at least to the extent described, in future haying operations on the place, but, as in the past, haying off the place will involve some use of tractor power.

While we feel that considerable progress has been made in bringing tillable ground back to fertility and crop production in the last two years, it is the woods work in the winter time which has generated most of the farm income, and in this the horses have also played a vital part. Last year two of us cut and brought out with the horses some 85 cords of pulpwood between early December and early April. In addition, we spent a week on a salvage cut of pine logs representing trees which were dead or dying. We also spent a week removing a blowdown of spruce trees some eight miles away from the farm. We rigged a loading ramp for the one-ton truck and took the horses down and back every day. We heat (and cook, in the wintertime) entirely with wood, and it is with the team that we get the next year's supply out during the winter.

We feel that the horses give us a great deal more effective power and flexibility in woods operations and for a much lower cost than would be the case with a farm tractor. Farm tractors are dangerous in the woods unless rigged with a protective frame over the driver; they cannot match horses for traction or ability to negotiate rough terrain; and they tend to suffer damage in such use. The skidder has great traction and will take out great amounts of wood, but the small woodlot owner will think twice before he lays out $25,000 or $30,000 on a machine which will tear up his land and destroy much of the timber which should be left, and which will be expensive to keep in repair. Much has been said recently about the increasing value of wood as a resource. The efficient utilization of our woods resources can only occur if the small and medium sized woodlots are properly harvested and managed. I believe horses may have a part to play in this because the skidder operator depends on clear cutting large lots as the only way to achieve enough volume to meet the time payments on his machinery. As a matter of fact, we hope to offer our services and those of the team as an alternative to skidders in harvesting small and medium sized woodlots, and I fully believe that as the cost of machinery and the required fuel continue to go up, more and more horses will appear in the small and medium sized woodlots of Maine.

In the early part of the Winter, if there is no snow on which to use a woods sled, we use a scoot, a small sled with hardwood runners flexibly assembled so that it goes easily on bare ground and over stumps and rocks. With stakes, a good load of four-foot wood, pulp or firewood may be taken out this way. On the wood sled four foot wood may also be carried, and, with the removal of the body, logs may be rolled up lengthwise on the sled bunk and held with stakes, and chain and log binder for removal to the skidway or yard. A good sled load would be one or one and a half cords (they used to take two in the old days), and 500 or so board feet of logs (perhaps six or eight of them). In addition, when logging, the team may be unhitched from the sled and used separately or together to twitch the logs out—depending on the size of the logs involved.

Far from the roar of the skidder, a man has time to think a little while the horses pull the loaded sled toward the yard through the sunny woods, their heads tossing and their nostrils steaming in the wintry sunshine. How have we gotten so far from our heritage of using draft animals in the woods and on the farm? In part, at least, the answer has to be cheap energy. Machines can do the work and then stand silently rusting and unattended until used again. No one has to feed them, water them, brush them, or give them pieces of apple or lumps of sugar. To work with and care for a team is a continuing, and to me, rewarding commitment, one that cheap energy has enabled fewer and fewer to enjoy. Whether we have gained anything in the process is of course a matter of opinion.

Now that the future of cheap energy seems uncertain, is there anything useful to be derived from our heritage of using draft animals? This depends a lot on your economic situation. For those who, like us, have a farm and woodlot and thus can use the horses productively year round, and at the same time raise their feed, such a course is most likely to make sense. If you should decide to try horses you will find that some of us greenhorns have taken the plunge, and that we, as well as some of the oldtime teamsters, will be glad to help you all we can.

Index

Edited by David Gauthier